Revision Notes
on
Plane Surveying

Revision Notes
on
Plane Surveying

W.S. WHYTE
A.R.I.C.S., A.I.A.S., F.R.S.A.

Principal Lecturer, School of Building, Surveying and Land Economy,
The City of Leicester Polytechnic

LONDON
NEWNES – BUTTERWORTHS

THE BUTTERWORTH GROUP

ENGLAND

BUTTERWORTH & CO. (PUBLISHERS) LTD.
LONDON: 88 Kingsway, WC2B 6AB

AUSTRALIA

BUTTERWORTH & CO. (AUSTRALIA) LTD.
SYDNEY: 586 Pacific Highway, 2067
MELBOURNE: 343 Little Collins Street, 3000
BRISBANE: 240 Queen Street, 4000

CANADA

BUTTERWORTH & CO. (CANADA) LTD.
TORONTO: 14 Curity Avenue, 374

NEW ZEALAND

BUTTERWORTH & CO. (NEW ZEALAND) LTD.
WELLINGTON: 26–28 Waring Taylor Street, 1
AUCKLAND: 35 High Street, 1

SOUTH AFRICA

BUTTERWORTH & CO. (SOUTH AFRICA) (PTY.) LTD.
DURBAN: 152–154 Gale Street

First published in 1971 by Newnes-Butterworths,
an imprint of the Butterworth Group

ISBN 0 408 00067 8

Printed by photo–lithography and made in Great Britain
at the Pitman Press, Bath

Preface

This book is intended to provide a set of brief revision notes covering the main aspects of plane surveying. It is not a textbook, the reader being assumed to have had recourse to the normal texts in his formal study of the subject, and this is reflected in the arrangement of the contents.

This book should be suitable for all students of plane surveying preparing for examinations, whether in universities, polytechnics, technical colleges or following correspondence courses. The content is appropriate to the relevant degree, diploma, HND, HNC and professional examinations in plane surveying in the fields of architecture, construction, engineering, estate management, planning and quantity surveying.

I am indebted to Messrs. Wild Heerbrugg (U.K.) Ltd., and Rank Precision Industries Ltd. for the use of illustrations of their theodolite reading systems. My thanks are also due to my colleague, Mr. R. Thomas, B.Sc., Lecturer in Mathematics and Computing in The City of Leicester Polytechnic, for his able assistance.

Leicester   W.S. WHYTE
1971

Contents

1 Introduction

Surveying operations are directed at two main purposes:

    (i)   The determination of the relative positions of points (natural or artificial features) on the surface of the earth so that they may be correctly represented on maps or plans, and

    (ii)   the setting-out on the ground of the positions of proposed construction or engineering works.

1·1 Branches of surveying

Geodetic Surveying is concerned with determining the size and shape of the earth and it also provides a high-accuracy framework for the control of lower-order surveys. The highest standards of accuracy are necessary and due allowance must be made for the curvature of the earth's surface.

Plane Surveying deals with areas of limited extent and it is assumed that the earth's surface is plane, hence no corrections are necessary for the earth's curvature.

Topographical Surveying is concerned with the measurement and mapping of the physical features of the earth.

Cadastral Surveying is concerned with the measurement, definition and mapping and recording of property boundaries.

Engineering Surveying covers surveys carried out as part of the preparation for, or execution of, engineering works.

1·2 Principles of survey practice

(i) Control — Every survey must first be provided with a high-accuracy framework, the later lower-order work then being fitted and adjusted to the framework. The traditional expression is 'work from the whole to the part'.

(ii) Economy of accuracy — The standard of accuracy aimed at should be appropriate to the needs of the particular task, and no higher. As a general rule, the higher the standards of accuracy aimed at, the higher the cost in time and money.

(iii) Consistency — The relative standards of accuracy of the various classes of work (control framework, detail fill-in) should be properly co-ordinated with one another and these standards should be maintained through the whole area of the survey.

(iv) Independent checks — Every survey and computation operation should either be self-checking or should be provided with an independent check.

(v) Revision — Surveys should, where possible, be planned in such a way that their later revision or extension may be carried out easily.

(vi) Safeguarding — The results of survey work should be available for use by other surveyors at a later date, thus all methods of working should be such as will be readily understood by other surveyors, and records should be properly maintained.

1·3   Survey methods

Triangulation – Framework of triangles in which all angles are measured but only one side length measured. The remaining sides are computed by solving triangles by sine rule.
Applications – (a) Geodetic and lower-order survey frameworks for control, angles measured by theodolite and one base side measured by taping or electromagnetic distance-measuring methods.
(b) Plane table survey, angles established graphically.

Trilateration – Framework of triangles in which all side lengths are measured.
Applications – (a) Geodetic control frameworks, lines measured by e.d.m.
(b) Detail survey, lines measured by chain, tape, or band – More conventionally known then as 'chain survey'.

Traverse – Chain of straight lines, connected at their ends, the length of each line and the angles between successive lines measured.
Application – Control surveys of any order. Angles by theodolite or compass or plane table. Lines by direct, optical or e.d.m. methods.

Radiation – Position fixing by an observed direction in azimuth and a measured distance. Also termed 'polar co-ordinates'.
Application – Detail survey from pre-fixed control points, with distance by direct, optical or electro-magnetic methods.

Offsetting – Position fixing by measurement of a distance at right angles to a control line.
Application – Detail survey – similar to use of plane rectangular co-ordinates.

1·4   Measurement and errors

Direct measurement – one made directly on the quantity being determined.

Indirect measurement – one deduced from measurement of other quantities.

True value of a quantity – never known, as regards angles and lengths, since all measurements subject to a variety of unknown errors.

Error – difference between true and measured values of a quantity.

Accuracy – conformity of a measurement to true value. To specify, the estimated error in the measurement must be found.

Accuracy of a measurement – specified by stating relative error, or ratio of error to measured quantity, thus giving degree of refinement obtained in the measurement.

Precision of a measurement – degree of refinement used in making the measurement.

Repeated measurement – accuracy of measured quantity obtained by taking several measures and meaning (averaging) the results obtained.

Precision of a set of measurements – degree of agreement between the measurements.

Most probable value (m.p.v.) – arithmetic mean of several independent and equally reliable measures of the particular quantity.

Weight of an observation – an indication of its reliability relative to other observations, usually expressed as a number.

Residual, or residual error – difference between m.p.v. of a quantity and an actual observed value.

2

1·4·1   Types of errors

Three classes of error:
- (a)   gross errors, or mistakes, or blunders
- (b)   systematic errors, including constant errors
- (c)   accidental errors, or random errors.

Gross errors — result from carelessness or inattention, work must be arranged to provide checks which will detect and eliminate these.

Systematic errors — always occur under same set of conditions. Cumulative, may be positive or negative. Equipment and methods should be arranged and used in such a way as to detect and eliminate these.

Accidental errors — the unavoidable small inaccuracies every measurement is subject to. Irregular in effect, positive or negative, cannot be eliminated.

1·4·2   Accidental or random errors

If gross and systematic errors are eliminated by appropriate work methods, only random errors remain to be allowed for. Random errors cannot be eliminated, they are subject to chance, their occurrence is then assumed to follow the laws of probability. These laws indicate, as regards repeated direct measurement, that:
- (a)   small errors occur more often than large ones
- (b)   positive and negative errors are equally likely, and
- (c)   very large errors seldom occur.

Real errors never being known, the residuals of the several observations in a set are treated instead. If the residuals of a large set of observations of a quantity are plotted as a frequency polygon, the result approximates closely to the Normal Distribution (or Gaussian) curve shown.

$$\text{Equation:}\quad y = \pi^{-1/2}\, h\, \exp(-h^2\, x^2)$$

Where h is an indication of the precision of the set.

Area under curve represents probability of all errors occurring (1). Number of errors in an interval proportional to area between the two ordinates at the ends of the interval.

If no systematic error, curve will be approximately symmetrical about the y axis, i.e. about arithmetic mean.

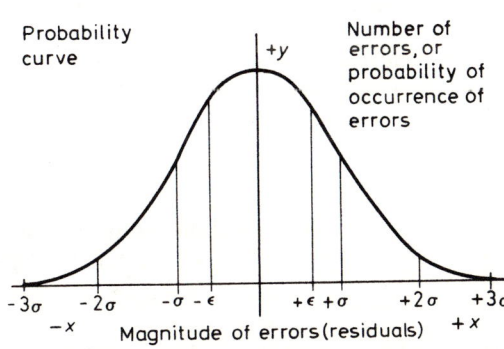

Principle of least squares — The sum of the squares of the residuals derived from the arithmetic mean is a minimum. Then

$$[v^2] = v_1^2 + v_2^2 + v_3^2 + \cdots v_n^2 = \text{a minimum, using } v \text{ as residual symbol.}$$

Average error of set of observations of a single quantity — The mean of all the errors (residuals) taken regardless of sign. Also termed mean error or mean deviation, symbol $\eta - 0.7979\,\sigma$, but see later.

Standard deviation of a single observation in a set is the square root of the sum of the squares of the residuals divided by the number of surplus observations. Also known as standard error, s.e., or mean square error, m.s.e., of the single observation. Symbol σ, sometimes m. Then

$$m = \sigma = \pm \left\{ [v^2]/(n-1) \right\}^{1/2} \text{ where } n \text{ is number of measures.}$$

Alternative notation

$$\sigma = \pm \left\{ \Sigma(x - \bar{x})^2/(n-1) \right\}^{1/2}, \text{ where } \bar{x} \text{ is arithmetic mean and } x \text{ the observation.}$$

If error distribution follows the normal distribution curve, then:
- 68% of all observations lie within one standard deviation of the arithmetic mean,
- 95% lie within two standard deviations of the arithmetic mean, and
- 99.7% lie within three standard deviations of the arithmetic mean.

Thus standard deviation is an indication of the reliability of observations, and probability of a residual exceeding 3 x standard deviation is so low that such a residual is suspect and should possibly be rejected.

Standard deviation or standard error of arithmetic mean of a set is equal to $\pm \left\{ [v^2]/n(n-1) \right\}^{1/2}$, i.e. s.d. of single observation divided by root of total number of observations. Indicates the reliability of the arithmetic mean, there being a 68% chance that the true value of the quantity will lie within $\pm$ one s.d. of the arithmetic mean. Symbol σ_{mean}, or σ_0, or m' or M. Then

$$\sigma_{\text{mean}} = \sigma_0 = \sigma/n^{1/2}, \text{ or } m' = m/n^{1/2}$$

Probable error divides the area under the curve into two equal parts, thus there are as many errors greater than the probable error as there are less than it. Also termed 50% error. P.E. equal to $0.6745\,\sigma$, symbols used p.e., ϵ, p, or r.
Then p.e. $= 0.6745\,\sigma \approx \frac{2}{3}\,\sigma$.
Probable error of mean is equal to $0.6745\,\sigma/n^{1/2}$.
Probable error little used today, misleading term since most probable error in a set is always zero.

Variance of a set of observations is the square of the s.d. of a single observation, it is an indication of the spread of results. Variance $= [v^2]/(n-1)$.

Arithmetic mean of a set of observation is simple average of the set, $\bar{x} = [x]/n$, where $\bar{x}$ is a.m., x the individual observation, and n the total number of observations in the set.
Rapid calculation method

- (a) select datum value *lower* than the a.m.,
- (b) total all residuals from the datum value and divide by n,
- (c) add the result to the datum value, this gives arithmetic mean.

Thus $\bar{x} = x_d + \Sigma(x - x_d)/n$, where x_d is the datum value,
or $\bar{x} = x_d + [x - x_d]/n$, using alternative notation.

Standard deviation may also be calculated by speedier method.

$$\text{Thus } \sigma = \pm \left\{ \Sigma(x - \bar{x})^2/(n-1) \right\}^{1/2}, \text{ but}$$
$$\Sigma(x - \bar{x})^2 = \Sigma x^2 - (\Sigma x)^2/n$$
$$\therefore \sigma = \pm \left\{ (\Sigma x^2 - (\Sigma x)^2/n)/(n-1) \right\}^{1/2}$$
$$= \pm \left\{ (n\Sigma x^2 - (\Sigma x)^2)/n(n-1) \right\}^{1/2}$$

Standard error formulae summary — *see* section 1.8.

4

Weight of an observation is an estimate of its reliability as compared with other observations. Better the observation greater its weight. Observation weight inversely proportional to square of its standard error, thus useful in comparing reliability of several observations.

Weighted mean of several varyingly weighted observations is the a.m. of all the observations *reduced to the same standard*. If the n observations of quantity x are respectively $x_1, x_2, \ldots x_n$, weights being p_1, $p_2, p_3, \ldots p_n$, weighted mean is then $x_0 = (p_1 x_1 + p_2 x_2 \ldots p_n x_n)/(p_1 + p_2 \ldots p_n) = [px]/[p]$.

Standard error of weighted mean is $\sigma_0 = \pm \{ [pv^2]/[p] (n-1) \}^{1/2}$.

Standard error of single observation of unit weight is $\sigma = \pm \{ [pv^2]/(n-1) \}^{1/2}$.

Standard error of single observation of weight p_r is $\sigma_r = \pm \{ [pv^2]/p_r(n-1) \}^{1/2}$.

Assigning weights to observations. Preferably weight observations in inverse proportion to the squares of their standard errors. If must be weighted before standard errors known, allot the weights according to relative precision of equipment used or the number of observations made.

1·5   Geometry

The following theorems are much used in survey work:

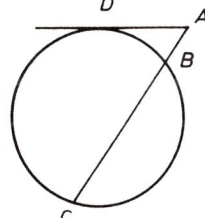

$AD^2 = AB.AC$
Tangent and secant

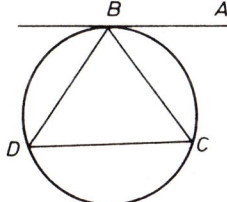

$A\hat{B}C = B\hat{D}C$
Chord and tangent

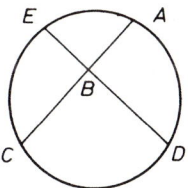

$AB.BC = EB.BD$
Intersecting chords

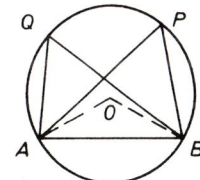

$A\hat{P}B = A\hat{Q}B = A\hat{O}B/2$
Chord and subtended angles

1·6   Trigonometry

The following sections cover the important functions and formulae of plane survey.

1·6·1   *Trigonometrical functions* (Assume circle radius $R = 1$)

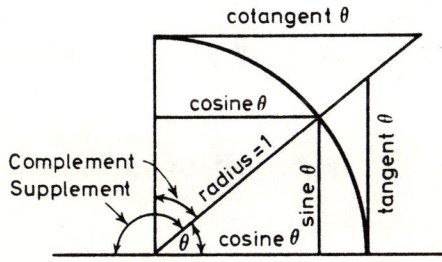

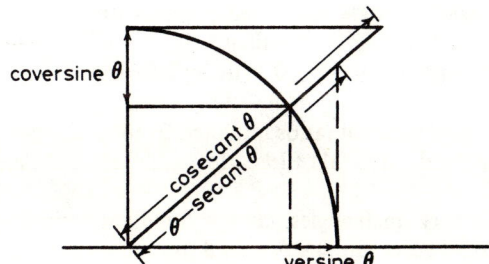

$$\text{cosecant } \theta = 1/\sin \theta$$
$$\text{secant } \theta = 1/\cos \theta$$
$$\text{exsecant } \theta = \sec \theta - 1$$
$$\text{versine } \theta = 1 - \cos \theta$$
$$\text{coversin } \theta = 1 - \sin \theta$$

$$\sin^2 \theta = 1 - \cos^2 \theta$$
$$\cos^2 \theta = 1 - \sin^2 \theta$$
$$\tan^2 \theta = \sec^2 \theta - 1$$
$$\cot^2 \theta = \operatorname{cosec}^2 \theta - 1$$

$$\sin \theta = 1/\surd(1 + \cot^2 \theta)$$
$$\cos \theta = 1/\surd(1 + \tan^2 \theta)$$
$$\tan \theta = \sin \theta/\surd(1 - \sin^2 \theta)$$

1·6·2 Sums and differences of two angles

$$\sin (A \pm B) = \sin A \cos B \pm \cos A \sin B$$
$$\cos (A \pm B) = \cos A \cos B \mp \sin A \sin B$$
$$\tan (A \pm B) = \frac{(\tan A \pm \tan B)}{(1 \mp \tan A \tan B)}$$

$$\sin A + \sin B = 2 \sin \{ (A + B)/2 \} \cos \{ (A - B)/2 \}$$
$$\sin A - \sin B = 2 \sin \{ (A - B)/2 \} \cos \{ (A + B)/2 \}$$
$$\cos A + \cos B = 2 \cos \{ (A + B)/2 \} \cos \{ (A - B)/2 \}$$
$$\cos A - \cos B = -2 \sin \{ (A + B)/2 \} \sin \{ (A - B)/2 \}$$

$$\sin (A + B) + \sin (A - B) = 2 \sin A \cos B$$
$$\sin (A + B) - \sin (A - B) = 2 \cos A \sin B$$
$$\cos (A + B) + \cos (A - B) = 2 \cos A \cos B$$
$$\cos (A + B) - \cos (A - B) = -2 \sin A \sin B$$

$$\sin 2A = 2 \sin A \cos A$$
$$\tan 2A = 2 \tan A/(1 - \tan^2 A)$$
$$\cos 2A = \cos^2 A - \sin^2 A$$
$$= 1 - 2 \sin^2 A$$
$$= 2 \cos^2 A - 1$$

1·6·3 Triangle solution formulae

Sides a, b, c. Angles A, B, C. $A + B + C = 180°$. $2s = a + b + c$.

Given three angles and a side. – Sine rule, $a/\sin A = b/\sin B = c/\sin C = 2R$, where R is radius of circumscribed circle. $b = a \sin B \operatorname{cosec} A$, $c = a \sin C \operatorname{cosec} A$.

Given three sides. – Cosine rule, $\cos A = (b^2 + c^2 - a^2)/2\,bc$. Alternatives are
$\sin A = (2/bc)\surd \{ s(s - a)(s - b)(s - c) \}$ or half angle formulae,
$\sin A/2 = \surd \{ (s - b)(s - c)/bc \}$
$\cos A/2 = \surd \{ s(s - a)/bc \}$
$\tan A/2 = \surd \{ (s - b)(s - c)/s(s - a) \}$
Triangle area $= \surd \{ s(s - a)(s - b)(s - c) \} = \frac{1}{2} ab \sin C$.

Given two sides and included angle. Cosine rule as above, $c^2 = a^2 + b^2 - 2\,ab \cos C$. Alternative, Napier's tangent rule, $\tan (A - B)/2 = \dfrac{a - b}{a + b} \tan (A + B)/2$ is preferable.

Given two sides and angle opposite one of them. – Sine rule, beware ambiguous case.

1·6·4 Circular measure and small angles

Radian – angle subtended at centre of circle by arc of length equal to circle radius.
2π radians $= 360°$. 1 radian $= 57° \; 17' \; 45'' = 206\,265$ seconds (approximately). θ is the usual symbol for angle in radians. $\theta = $ arc length/radius. Arc $= r\theta$.

Trigonometrical ratios obtained from
$\sin \theta = \theta - (\theta^3/3!) + (\theta^5/5!) - (\theta^7/7!) + \ldots$, and $\cos \theta = 1 - (\theta^2/2!) + (\theta^4/4!) - (\theta^6/6!) + \ldots$

For very small angles, $\tan \theta \approx \theta$ radians $\approx \sin \theta$.
To 5 figures, $\tan \theta = \theta = \sin \theta$, up to $2°$.
To 4 figures, $\theta = \sin \theta$, up to $5°$, and $\theta = \tan \theta$, up to $4°$.

1·7 Measurement and computation

General hints:
- (a) record all computations in ink
- (b) make alterations by crossing out and re-writing above — never erase
- (c) group figures in threes in either direction from the decimal point, no commas
- (d) arrange figures in columns
- (e) use self-checking form of calculation, alternatively arrange independent check calculation from the original data
- (f) arrange calculations so that they 'read' like a piece of written matter, complete with punctuation
- (g) select formulae appropriate to the calculation method, e.g. $a \times b \times c$ is suited to logs, but $(a + b + c)(d + e)$ is not
- (h) work generally to one more place of decimals than is required in the answer, round-off as required after completion.

1·7·1 Significant figures

If readings are estimated to nearest division on a graduated scale, the maximum estimation error will be ± ½ division, final recorded digit being subject to a maximum error of half its value.

Number of significant figures in a product should be same as number of significant figures in the least significant factor, or greater by one significant figure. General rule: 'retain one more significant figure in the product than in the least significant factor'.

Same rule for division and powers. For roots, number of significant figures in root should be equal to number of significant figures in the radicand, or it should be greater by one.

If in doubt, apply following rules:
If $P = x \cdot y$, $\delta P = P\{(\delta x/x) + (\delta y/y)\}$ If $Q = x/y$, $\delta Q = Q\{(\delta x/x) + (\delta y/y)\}$
If $R = x^n$, $\delta R = Rn\delta x/x$ If $R = \sqrt[n]{x}$, $\delta R = R\delta x/nx$
(δx represents uncertainty or inaccuracy in x.)

1·7·2 Number of figures in natural and log trigonometrical function

	0·1"	1"	10"	1'	1°
Angle read to limit of					
Number figures required in log functions	7	6	5	4	4
Decimals required in natural functions	7	6	5	4	4

1·7·3 Accuracy of logarithms

Assuming error of 4 in the last figure of a logarithm:

Number of figures in logarithm	4	5	6	7
Proportional fractional error	1/1000	1/10 000	1/100 000	1/1 000 000

1·7·4 Constants and conversions

$\pi = 3 \cdot 141\ 592\ 6536$ 1 radian $= 57° \cdot 295\ 779\ 5130 = 206\ 264" \cdot 8062$
$e = 2 \cdot 718\ 281\ 8285$ cosec $1" = 206\ 264 \cdot 8062$

1 grade $= 0 \cdot 9° = 54' = 3240"$. $1° = 1 \cdot 111\ 111\ 1111$ grades.

1 metre $= 3 \cdot 280\ 84$ ft.   1 ft $= 0 \cdot 3048$ metres.
1 km $= 0 \cdot 621\ 371$ mile.   1 mile $= 1 \cdot 609\ 344$ km.

1·8 Summary of standard error formulae

Quantity

Single observation in a set

$$\sigma = \pm \left\{ [v^2]/(n-1) \right\}^{1/2}$$

Mean of n observations

$$\sigma_0 = \pm \left\{ [v^2]/n(n-1) \right\}^{1/2}$$

Multiple of quantity, such as $X = a \times x$, where a is a number and x a quantity having standard error of σ.

$$\sigma_X = \pm a.\sigma$$

Sum or difference of several quantities, such as $X = x_1 + x_2 - x_3 + \ldots x_n$, the quantities having standard errors of $\sigma_1, \sigma_2, \ldots \sigma_n$.

$$\sigma_X = \pm (\sigma_1^2 + \sigma_2^2 + \sigma_3^2 + \ldots \sigma_n^2)^{1/2}$$

If $\sigma_1 = \sigma_2 = \sigma_3 = \ldots \sigma_n = \sigma$, then

$$\sigma_X = \pm \sigma.n^{1/2}$$

Sum or difference of multiples of several quantities, such as $X = a_1 x_1 + a_2 x_2 - a_3 x_3 + \ldots a_n x_n$, the quantities $x_1, x_2, \ldots x_n$, having standard errors of $\sigma_1, \sigma_2, \ldots \sigma_n$.

$$\sigma_X = \pm \left\{ (a_1 \sigma_1)^2 + (a_2 \sigma_2)^2 + \ldots (a_n \sigma_n)^2 \right\}^{1/2}$$

Product of quantities, such as $X = xyz$, these having standard errors of $\sigma_1, \sigma_2, \sigma_3 \ldots \sigma_n$.

$$\sigma_X = \pm (xyz) \left\{ (\sigma_x/x)^2 + (\sigma_y/y)^2 + (\sigma_z/z)^2 + \ldots \right\}^{1/2}$$

Function of several quantities, such as $X = f(x_1, x_2, x_3, \ldots x_n)$, these having standard errors of $\sigma_1, \sigma_2, \ldots \sigma_n$.

$$\sigma_X = \pm \left\{ \left(\frac{\partial X}{\partial x_1} \sigma_1 \right)^2 + \left(\frac{\partial X}{\partial x_2} \sigma_2 \right)^2 + \ldots \left(\frac{\partial X}{\partial x_n} \sigma_n \right)^2 \right\}^{1/2}$$

2 Direct Linear Measurement

This chapter deals with direct linear measurement using the land chain, the steel band chain or the various types of steel tapes.

2·1 Corrections to measured lengths

All directly measured distances require the application of certain corrections to the observed measurements. The particular corrections required depend upon the accuracy demanded of the measurement and the equipment used. The accuracy of a linear measurement is stated as the ratio of the error at the end of the line being measured to the length of the line.

2·1·1 Correction for slope

All measurements must be reduced to the horizontal. Generally measurements are made on the ground surface (or parallel to it) then the ground slope determined. Slope may be measured by (a) the vertical angle, or (b) difference in height between the ends of the line.

Where slope angle measured, α, and measured slope distance l, correction to reduce to horizontal is $- l(1 - \cos\alpha) = - l$ versine α.

Where difference in height of line ends d, measured, and measured slope distance l, correction is $- d^2/2l$. For greater accuracy, use $- (d^2/2l + d^4/8l^3)$. Using the first term only gives relative accuracies of 1/1 000 000 at 1 in 20 gradient, 1/30 000 at 1 in 8, 1/2000 at 1 in 4.

2·1·2 Correction for standardisation

For a tape of nominal length l, and actual length $l \pm \delta l$, as obtained by comparison with a standard, the error per unit length is $\pm \delta l/l$. Where the length of a line is measured as d_m, the *true* length being d_t,

then $d_t = d_m \pm d_m \, \delta l/l = d_m (1 \pm \delta l/l)$.

Correction $= \pm d_m \, \delta l/l$.

If tape too long, correction positive,

If tape too short, correction negative.

2·1·3 Correction for temperature

Where a line length is measured as l, the measured field temperature t_f, the temperature at which the tape was standardised being t_s, and the coefficient of linear expansion of the tape metal is c, correction to measured length $= \pm lc(t_f - t_s)$.
(An average value of c for steel may be taken as $11 \cdot 2 \times 10^{-6}$ per $1°$C.)

2·1·4 Combined correction for standardisation and temperature

The above two corrections may be combined into one if the temperature at which the tape will be equal in length to its nominal length is first determined.

Tape nominal length l, actual length at standardisation temperature $l \pm \delta l$, standardisation temperature t_s. The actual tape length will be equal to the nominal length at a temperature t_s', difference between t_s and t_s' being t degrees and the coefficient of linear expansion c.

Standard length at $t_s = (l \pm \delta l)$.
Required to reduce (or increase) tape length by δl, and $\delta l = (l \pm \delta l) ct$.
Then $t = \delta l/c(l \pm \delta l)$.

$$t'_s = t_s \mp \delta l/(l \pm \delta l)c \simeq t_s \mp \frac{\delta l}{lc}$$

Finally, correct measured line length for a temperature change of $t'_s - t_f$.

2·1·5 Correction for tension

Line length measured as l, applied tape tension T_f, the tape being standard at tension T_s. Cross-sectional area of tape A, Young's modulus of elasticity E, correction to measured length $= \pm l(T_f - T_s)/AE$.

(Units must be compatible i.e. l and correction in metres, T in newtons, A in mm^2, and E may be taken as 20×10^4 N/mm^2 for steel and $14·5 \times 10^4$ N/mm^2 for Invar.)

2·1·6 Correction for sag

Tape standardised on the flat:
If used on the flat, no correction required.

If used in catenary, correction $= -w^2 l^3/24\, T^2$, or $-W^2 l/24\, T^2$, where w = tape weight per unit length, W = total tape weight, l = tape length, and T = applied tension. W and T must be in the same units, either both in newtons or both in kilogramme force.
(This would be followed by a correction for tension if T_f not equal to T_s.)

If used in catenary, *and the tape ends are not level,*
correction $= -w^2 l^3 \cos^2 \alpha/24\, T^2$, or $-W^2 l \cos^2 \alpha/24\, T^2$, where α = slope angle.

Tape standardised in catenary:
If used on the flat, apply an *additive* sag correction, using standard tension value.
Correction $+ W^2 l/24\, T^2$.

(If applied field tension not the same as standardisation tension, apply a further $\pm$ correction for the difference in tension, $T_f - T_s$.)

If used in catenary, with standard tension, *no correction.*

If used in catenary, with different tension from standard, apply an 'additive' sag correction to get 'flat' length, then a negative sag correction based on T_f, the applied tension.
(This would be followed by a correction for tension if T_f not equal to T_s.)

| Standardised in catenary | Used on flat | 'Modified' catenary |

Field determination of tape weight:
If y = mid-point sag, w = weight per unit length of tape, T = applied tension, and l = tape length along curve between ends,

$$w = 8\, Ty/l^2.$$

10

2·1·7 Correction for altitude

In higher-accuracy work, measured lengths must be reduced to the equivalent distance at mean sea level.

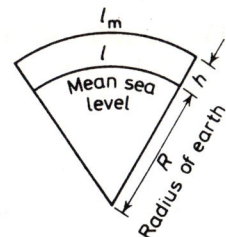

Line length measured as l_m, at an altitude of h above m.s.l., correction to reduce to m.s.l. is $-l_m h/R$.

R may be taken as 6370 km, then correction $= -l_m h/(6370 \times 10^3)$ metres. This reduces to approximately $15.7\,h \times 10^{-6}$ metres per 100 metres distance.

2·2 Sources and effects of errors in measurement and correction

2·2·1 Tape standardisation

Errors in tape length should be detected by comparing with a standard.
Example: An error of 1 mm in the length of a 30 m tape results in a proportional error of 1/30 000.

2·2·2 Alignment

If a tape of length l lies straight, with one end on line and the other off line by a distance d, error in measured length $= d^2/2\,l$.
Example: 30 m tape, one end 0·1 m off line, prop. error approx. 1/200 000.

2·2·3 Straightening

If the two ends of a tape of length l lie on line, but the centre is d off line, the two halves lying straight, error in measured lengh $= 2d^2/l$.
Example: 30 m tape, centre 0·1 m off line, prop. error approx. 1/50 000.

2·2·4 Reading the tape

For a tape graduated in millimetres, probable reading error of one end ± 0.2 mm.
Considering both ends, ± 0.3 mm.
The prop. error in a 30 m tape length is approx. 1/100 000.

For a tape graduated in centimetres, probable reading error of one end ± 1 mm, considering both ends, ± 1.4 mm.
The prop. error in a 30 m tape length is approx. 1/20 000.

2·2·5 Marking the tape ends

Marking errors may amount to three times the reading error.
The values quoted for reading errors become 1/33 000 and 1/6600 approx.

2·2·6 Temperature measurement

Errors in temperature measurement have been found to range up to 3°C or more, depending on circumstances. An error of 3°C in temperature measurement will result in a proportional error of approx. 1/30 000.

2·2·7   Recording tension

For a 30 m steel tape, 13 mm width, a variation of 40 N in the recorded tension will result in a linear error of ± 3 mm, a proportional error of 1/10 000.

2·2·8   Sag determination

The accuracy of the correction for sag depends upon the accuracy of both the applied tension and the tape weight. (The effect of error in the measurement of the slope of the tape may be ignored.)
A 10% error in the determination of either the tension *or* the tape weight will result in a measurement proportional error of the order of 1/10 000.

2·2·9   Slope measurement

Error in determining height of ends
For a 30 m slope length, and a difference in height between the ends of 2 m, an error of ± 0·01 m in the determination of the height difference will result in a proportional error of approx. 1/40 000. In practice, levels may be determined to within ± 2 mm.

Error in measuring slope angle
An error of ± 20″ in the measurement of a slope angle of 10° will result in proportional error of approx. 1/60 000.
The greater the slope angle the more accurately it must be measured.

2·2·10   Reduction to mean sea level

The percentage error in the calculated correction is equal to the percentage error in the determination of the height above m.s.l.

2·3   Linear measurement methods

The choice of field methods and equipment depends upon the purpose of the measurement and the accuracy required.

2·3·1   Chaining with the land chain or steel band chain

Chain or band fully supported on the surface of the ground, no tension handles, no temperatures observed, slopes stepped or measured by clinometer, alignment by eye.
Corrections: standardisation; slope.
Accuracy: land chain 1/100 − 1/1000, steel band chain 1/1000 − 1/2000.

Application: chain survey and detail measurement.

2·3·2   Surface taping

Steel tape fully supported on ground surface, tension measured, temperatures measured, slopes by clinometer, alignment by eye.
Corrections: standardisation/temperature; tension; slope;
Accuracy: 1/1000 −1/20 000, depending on method and care taken.

Application: third-order traversing, minor base measurement, and higher-accuracy detail measurement.

2.3.3 Catenary taping

Three common methods used, according to requirements. All may be used for traversing and minor base measurement if accuracy suitable.

Method 1
Two chainmen hold tape ends at waist height, against a ranging rod for stability, end marking by plumb-bob, tension measured, temperatures measured, slopes by clinometer, alignment by eye.
Corrections: standardisation/temperature; sag (if necessary); tension; slope.
Accuracy: 1/1000–1/5000.

Method 2
Tape suspended over 1 m high stakes at its ends, otherwise as method 1, including the corrections necessary.
Accuracy: 1/1000–1/10 000.

Method 3
Tape suspended, one end held level with centre of end of theodolite horizontal axis, other end over a pre-driven stake. Otherwise as above, except that alignment and slope by theodolite.
Corrections: As for other methods, but reduction to m.s.l. may be required.
Accuracy: 1/5000–1/50 000.

2.3.4 Base line measurement

Measurement for geodetic base lines. Invar tapes in catenary using special equipment. All precautions. observed, all corrections applied.
Accuracy: 1/500 000 – 1/1000 000.

2.3.5 Simplified BRS taping method

The modified procedure was devised by the Building Research Station for improving the accuracy of steel tape measurement on construction sites. The principal sources of error in site work are (a) incorrect tension applied, and (b) the variation of temperature from standard. Assuming the use of a standard 30 m steel tape, 10 mm x 0.2 mm section, BRS suggest the following methods of correction, and the final corrected measure should have prop. error not exceeding 1/5000.

(a) Tension and sag
Whether the tape is used fully supported *or* in catenary, use a spring balance to apply a tension of *not more than* 70 N to the tape.
The resulting error in measurement will not exceed ± 3 mm, giving a maximum proportional error of 1/10 000. No corrections need be applied.

(b) Temperature
Measure air temperature by thermometer. Correct the measured length for temperature variation by deducting (or adding) 1 mm per 10 m of tape length per 10°C difference between air and standardisation temperatures.

For a fully supported tape it is suggested that the tape be supported on 25 mm wooden blocks to insulate it from ground temperature.
If the tape is exposed to strong sunlight, add 5°C to the observed air temperature.

This method should result in a proportional error of not more than 1/10 000.

Note: Slopes should be corrected for in the normal way.

2·4 Chain surveying

The figure shows layout of a typical chain survey of a small area. *Chain survey* means a survey using direct linear measurement only, no angle measurements

The *control framework* consists of a network of triangles formed by *chain lines,* those being measured by chain, tape or band.

A *detail line* is an extra line run close to detail, it is not an essential part of the control framework.

Detail is picked up by synthetic or steel tape *offsets* or *ties* from the framework lines.

An *offset* is a measurement at right angles to a chain or detail line.

A *tie* is one of a pair of measurements run from a pair of points on a chain or detail line to a point of detail.

A *check line* is an extra line which serves purely as a check on a framework triangle.

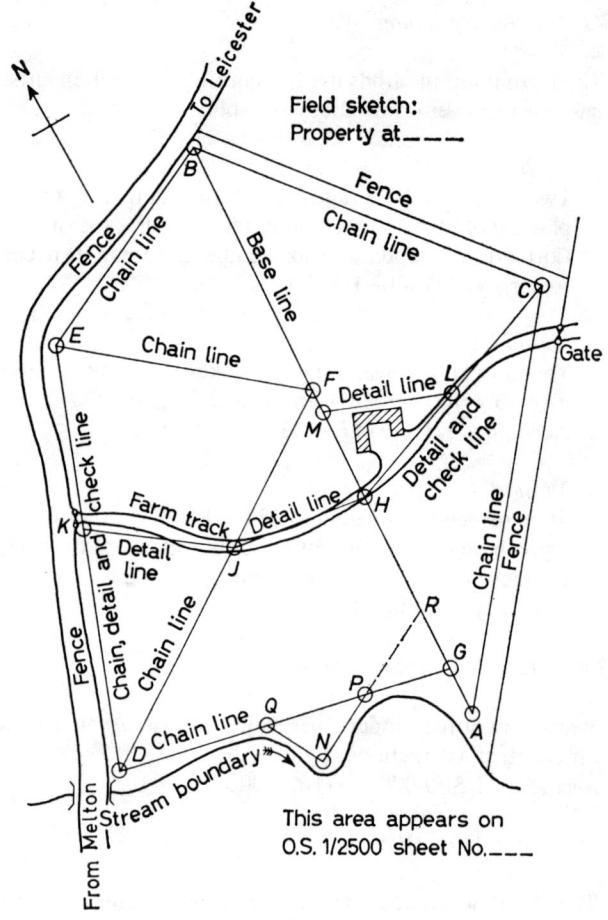

The *base line* of the framework is a long line, running through the whole survey area, and it serves to provide the base for all the triangles of the framework.

The *triangles* in the framework must be well conditioned, i.e. all angles must lie between 30° and 120°.

Each triangle must be provided with a *check measurement* in addition to the measured lengths of its three sides, in order to detect any error in the measurement of the sides.

Triangles may be checked or 'proved' by measuring any of the lines shown broken in the figures (a) to (d).

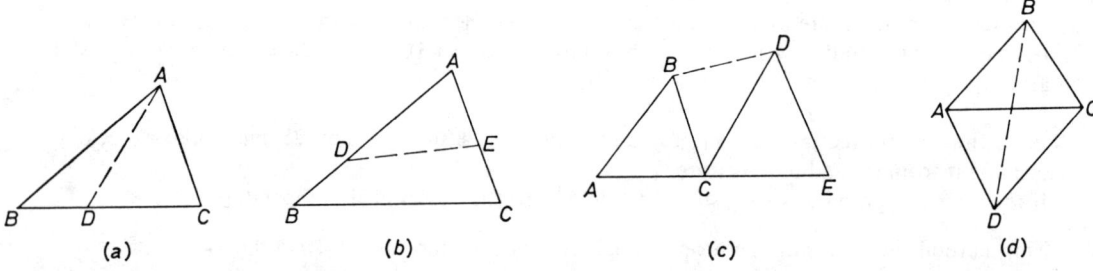

2·4·1 Rules for the selection of control framework lines

The base line should be as long as possible and preferably should run through the centre of the survey area.

All triangles should be based on the base line or tied back to it.

All triangles must be provided with checks.

Run lines as close to detail as possible to reduce offset lengths.

Run all lines over ground which is as clear of obstacles and as near horizontal as possible.

Keep the number of lines to the minimum provided the accuracy of the work is not impaired.

2·4·2 Booking chain surveys

Each chain, detail, or check line must be recorded separately in the field book, starting a new line on a new page.

To identify the lines:
> *method 1.* Give each station a letter of the alphabet, then 'DE' means 'line measured from Station D to station E'.
> *method 2.* Number each line and, on the preliminary sketch, draw an arrow alongside to indicate the direction of measurement.

The figure gives an example of the booking of one line of a chain survey.

The first page of the field book should contain a sketch of the framework (as on the previous page) together with the bearing of one line of the frame from north and details of surveyor, job name and address, weather, etc.

All measurements are noted in metres and decimals, generally to one place of decimals in chain survey, perhaps to two places where building details are concerned.

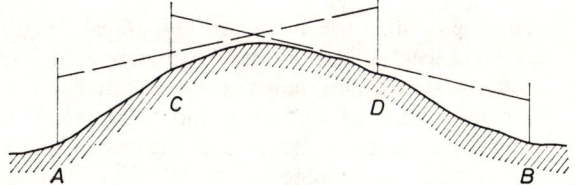

2·4·3 Obstacles to chaining or ranging

The following sections show some typical problems and solutions.

Ranging a line over a hill
Locate line ends, A and B, mark with rods. Select C and D, each side of crest, such that A and B are visible from both C and D.
Station chainmen at C and D.

Chainman at C lines rod D in with rod B.
Chainman at D then lines rod C in with rod A.

The procedure is repeated as necessary until A, C, D and B are in one straight line and no further movement possible.

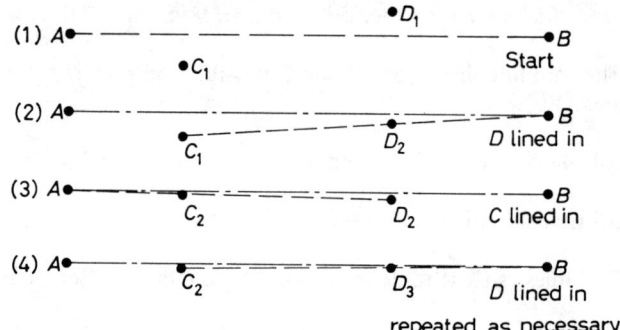

Obstacle prevents direct measurement but not ranging

(a) Obstruction cannot be chained around – Obstacle like a river.
(i) Line AB ranged over. Right angles set out at A and C, then at F, and D fixed. E lined in with D and B. Then BA = DF x AD/FE.
(ii) AC set out, C placed so that ∠ACB is 90°. DA set out, in line with and equal to CA. Right angle set out at D. EA = AB.

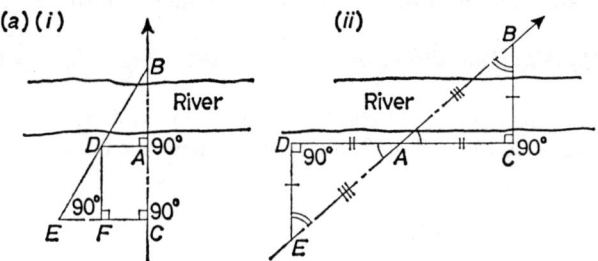

(b) Obstruction can be chained around – Obstacle such as a pond. All angles set out by linear measurement or box-sextant.

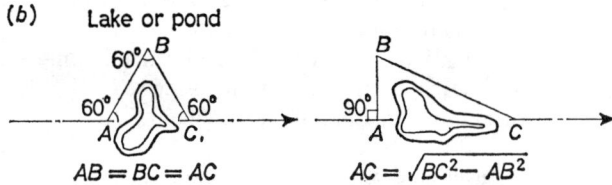

Obstacle prevents measurement and ranging

Typical examples are buildings which cannot be demolished. In the first figure, equal offsets placed at A and B, to locate C and D. The line CD ranged to E and F. Offsets set out at E and F, equal in length to AC and BD, then G and H located and line extended.

The second example is merely establishment of an equilateral triangle around the obstacle.

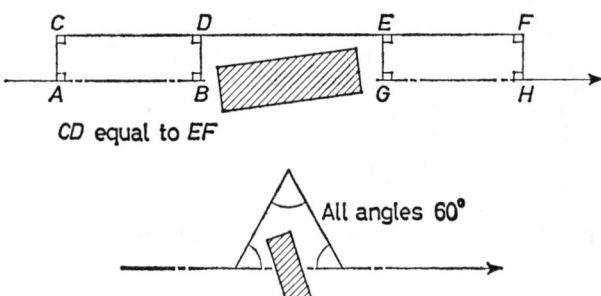

2·4·4   *Plotting chain surveys*

General procedure: Draw the base line to scale length, plot triangles by compasses and scale. Scale off check line lengths for verification, compare with field measured lengths.
Mark all station points, plot all detail line by line.
Add Northpoint and, where appropriate, Ordnance Survey Large Scale Map Number. Complete with firm's name, job name, surveyor, draughtsman, date, etc.
Add drawn scale if required.

3 The Level and the Theodolite

These are the principal instruments used in surveying. Both consist essentially of a telescope together with various rotational axis arrangements and associated levelling devices, while all theodolites and some levels also incorporate devices for reading directional angle measurements on graduated circles.

3·1   The surveying telescope

The simplest form of refracting telescope is a telescopic metal tube with a convex lens fixed in each end of the tube. One lens (the *objective*) is pointed at the distant object which is to be viewed, while the other serves as an eyepiece for viewing the image formed by the objective lens.

3·1·1   Convex objective lens

If a beam of light rays is passed through a convex lens, parallel to the principal axis of the lens, the light rays are refracted and converge to pass through the focal point of the lens, F.

Light rays which pass through the optical centre of the lens (O) will not be refracted.

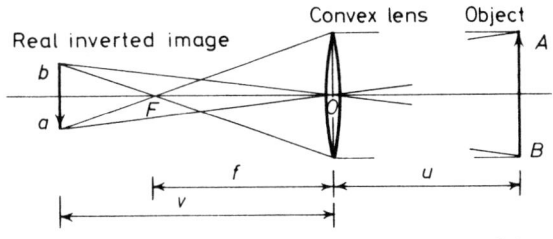

f = focal length ; *v* = image distance ; *u* = object distance

If a distant object is viewed, a real inverted image is formed at a distance *v* from the lens. The relationship between the lens focal length and the object and image distances is

$$1/u + 1/v = 1/f$$

The sign rules for lenses are:
  (i)   *f* is positive for a convex lens, negative for a concave lens
  (ii)   *u,* as measured starting from the lens, is negative if measured in the same direction as the light travels, positive if measured in the opposite direction to the light rays
  (iii) *v,* as measured starting from the lens, is positive if measured in the same direction as the light travels and negative if measured against it.

3·1·2   Convex eyepiece lens

A convex lens may be used as a magnifying glass *or* as an eyepiece for viewing the image formed by the objective lens of a telescope.

When used as an eyepiece, a magnified virtual image ($a_1 b_1$) is formed from the real image (ab) formed by the objective lens.

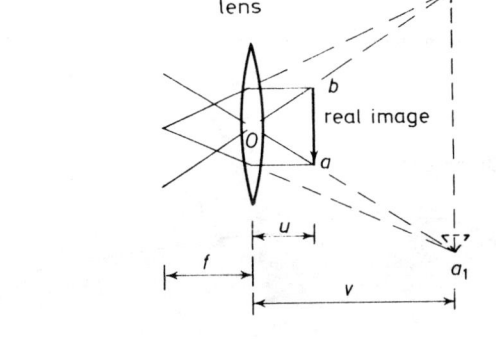

Magnification is

$$m = \frac{\text{size of image}}{\text{size of object}} \approx \frac{\text{image distance}}{\text{object distance}} = \frac{v}{u}$$

17

3·1·3   Lens defects

Chromatic aberration results from refracted white light being split into its component colours, making focusing difficult. This is overcome in the objective lens by using a compound of two lenses, a flint-glass concave lens cemented to a crown-glass convex lens, the compound acting as a single lens.

Spherical aberration results from the spherical lens surface refracting those incident rays near the lens edge more than those incident at the lens centre. Overcome by compounding two lenses (as above) or using two separate lenses whose aberrations are mutually cancelling.

In the *Ramsden Eyepiece* used on many survey telescopes, two identical plano-convex lenses (with curved faces facing) at a distance apart equal to 2/3 the focal length of either are used. The combination acts in the same way as the single convex lens eyepiece.

3·1·4   Kepler's telescope

This telescope is based on the principle of two convex lenses detailed above. In order to define a sight line, a diaphragm carrying a reticule of cross-lines (originally spider web) engraved on a glass plate is inserted at the focal plane of the eyepiece lens. To focus on an object the objective lens must be moved towards or away from the eyepiece until the image is formed at the diaphragm.

Parallax is the condition resulting when the objective image and the cross-hairs do not lie in the same plane — the eyepiece cannot be focused on both at once. To eliminate, focus eyepiece very carefully on the cross-hairs, then focus the objective carefully — check if eye movement results in the cross-hairs moving with respect to the object image, if so, re-focus eyepiece then objective.

The sight line defined by the centre of the cross-hairs and the optical centre of the objective lens is the *collimation line* of the instrument.

3·1·5   Galileo's telescope

This telescope uses a convex objective lens and a concave eyepiece lens. No real image is formed, only an erect virtual image, and the sight line cannot be precisely defined hence it is of no practical value in survey work.

3·1·6   Modern survey telescopes

Early surveying telescopes were on the Kepler principle, the two lens systems being fixed to separate sliding tubes in order to vary the distance v. These are known as *external focus* and are no longer used.

Modern telescopes have the distance between eyepiece and objective fixed, and the focusing is arranged by the movement of an extra diverging lens which is fitted about the centre of the telescope tube.

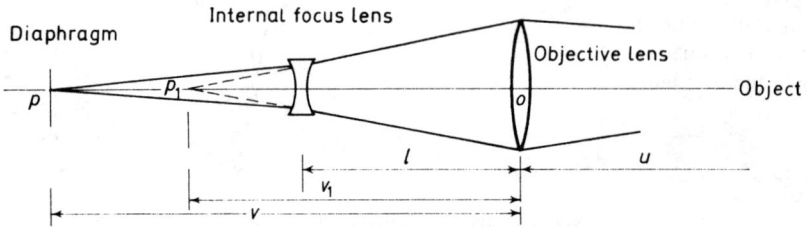

Without the internal focus lens the object image would be formed at P_1. The internal focus lens diverges the rays to form the image at P. Instead of moving the objective the internal focus lens is moved forward or back until the image appears at the cross-hairs at P.

Using the standard relationship, $1/u + 1/v_1 = 1/f$, $\therefore v_1 = fu/(u-f)$

If P_1 is regarded as a *virtual object* giving an image at P, and the focal length of the objective is f while that of the internal focus lens is f_1,
then

$$-1/(v_1 - l) + 1/(v - l) = -1/f_1, \quad \therefore f_1 = (v-l)(v_1 - l)/(v - v_1)$$

The *field of view* of a telescope is the proportion of the horizon that can be seen through the telescope. Specified as the angle subtended at the instrument centre by the visible field, it is, effectively, the angle subtended at the centre of the objective by the eyepiece, and it is fixed by the eyepiece diameter and the distance between the eyepiece and the objective. It is generally stated by manufacturers as the angle *or* as the linear width of field at a stated distance.

Magnification was considered earlier. For a telescope it can also be stated as the ratio of apparent field of view to actual field of view. If the magnification is increased, while the apparent field remains constant, the real field of view must decrease.

Image brightness depends upon the objective aperture diameter and the magnification. An increase in eyepiece magnification results in a diminution of the image brightness, and in poor observing conditions it may be of assistance if a lower-power eyepiece is used to give a smaller but brighter image.

Resolving power or *resolution* of a telescope is its ability to distinguish between directions to objects very close together at considerable distances. Resolving power of a telescope depends on the magnification, the image brightness and the quality and arrangement of the lenses used. The human eye can resolve objects with a separation of 1 or 2 minutes while typical telescope resolution is 3 or 4 seconds.

3.1.7 The external focus telescope and stadia measurement

The figure illustrates how the two horizontal stadia lines on the telescope diaphragm may be used for the measurement of distance. The points a, b, represent the stadia 'hairs' while A, B, are the equivalent points (points apparently cut by the stadia lines) on a distant staff held vertically at right angles to the telescope collimation line. The distance AB is the *staff intercept, s*, while i is the distance between the stadia lines.

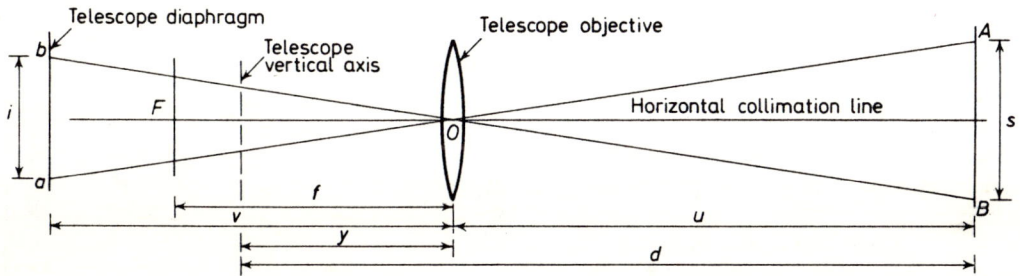

If the focal length of the objective lens is f, then

$$1/v + 1/u = 1/f$$

$$\therefore \quad u = (u/v)f + f$$

but $\quad u/v = AB/ab = s/i$

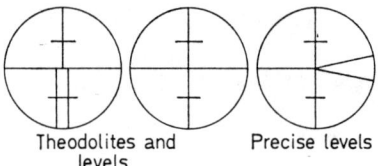

Theodolites and levels   Precise levels

$$\therefore \quad u = (f/i)s + f$$

Typical reticule patterns with stadia cross-hairs

and $\quad d = u + y$

$$= (f/i)s + (f + y)$$

The quantity f/i is a constant, $(f + y)$ varies since y changes as the objective is moved to focus the instrument, but $(f + y)$ is usually regarded as a constant and thus

$$d = Ks + c$$

Normally f/i is arranged to be exactly 100, and the instrument additive constant c is quoted by the manufacturer. If not known, both constants can be determined by several trial observations over a known measured distance, until simultaneous equations can be solved.

Porro introduced an extra *anallactic lens* into the telescope, with the object of eliminating the additive constant c (the telescope still being external focus, since the anallactic lens did not act as a focusing lens).

3·1·8   The internal focus telescope and stadia measurement

The modern internal focus telescope is not perfectly anallactic, but it is constructed in such a manner that the additive constant is so small that it may be ignored, and for all practical purposes $d = Ks$, the value fixed for K normally being 100.

3·2   The spirit level

Used on both levels and theodolites to determine level and verticality. Sealed glass vial, part filled with alcohol, ether or chloroform, the interior surface ground to a barrel shape with a constant radius of curvature in longitudinal section. *Principal tangent* of level tube is the tangent to the longitudinal curve at the centre of the vial. Vial usually graduated at 2 mm intervals.

3·2·1   Sensitivity of a level vial

Sensitivity of a level vial is the angle through which its principal tangent must be tilted to move the bubble along the tube one graduation interval. Sensitivity $= x = d/R$, where d = length of one tube graduation division and R = radius of inner surface of vial longitudinal section. $x'' = d \times 206265/R$.

Sighting on a distant staff, where a bubble shift of y graduations covers a staff intercept S, the angle of tilt being α, then

$$\alpha = S/D \text{ radians}$$

and

$$\alpha = dy/R \text{ radians}$$

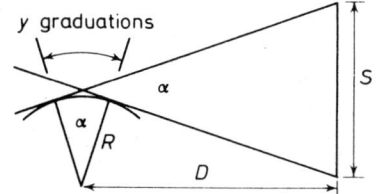

Therefore
$$dy/R = S/D$$
and
$$R = dDy/S.$$
but
$$x = d/R$$
$$= dS/dDy$$
$$= S/Dy \text{ radians}$$
$$x'' = S \times 206265/Dy \text{ seconds}$$

3·2·2   Coincidence reading bubble

Used on higher-quality tilting levels and often as the altitude bubble on theodolites, this spirit level is enclosed with a prism system which presents images of half of each end of the bubble. These images are viewed in an eyepiece and when they are coincident the principle tangent of the level tube is set horizontal. These allow more accurate levelling than the simple open bubble tube.

3·3   Circle reading mechanisms

Early theodolites used the vernier scale or the microscope micrometer for reading the circles. The latter is now obsolete, but many vernier theodolites are still in use and the system is used on some levels equipped with a horizontal degree circle. Most theodolites produced today are glass arc types with optical reading systems.

3·3·1   The vernier scale

A graduated scale placed alongside the reading index, as shown in simple form.

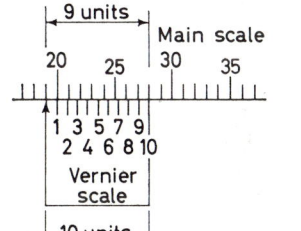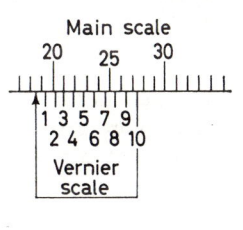

Number of divisions — In a direct vernier, if the vernier scale has n subdivisions, then it is made equal in length to $(n-1)$ sub-divisions of the main scale.

Least count of a vernier — The smallest unit the vernier will resolve, equal in value to the smallest sub-division of the main scale divided by the number of divisions on the vernier scale. If value of main scale sub-division is d, (minutes, seconds, millimetres, etc), least count = d/n.

Double vernier — One with two scales, one reading each way from the index. Used when the main scale can be read in either direction.

Extended vernier — One in which the n sub-divisions of the vernier scale are equal in length to $(2n-1)$ sub-divisions of the main scale. Read as a direct.

Retrograde vernier — One in which the n vernier sub-divisions are equal in length to $(n+1)$ main scale sub-divisions. This vernier scale reads in the *opposite* direction to the main scale. Read as a direct vernier.

To read a vernier — Note the main scale reading immediately before the index, then add the value of the vernier graduation which most nearly coincides with a main scale graduation.

Example — In the right-hand portion of the figure, the main scale reads 18. Vernier least count is 0·1. Coincident vernier graduation is number 3. Add 3 x 0·1, 0·3. Total reading, 18·3 units.

3.3.2   Optical reading systems

Used on theodolites (or levels) with glass circles. A beam of light (natural, or artificial for night working) is passed through the circle, the graduations being sharply silhouetted against the light, and these may be read by any of the following arrangements.

Circle microscope. The most basic system, the graduation's image merely being viewed through a microscope with a fixed index line in its field.

Example: Watts Microptic Transit, circles graduated at 5′, single minutes estimated.

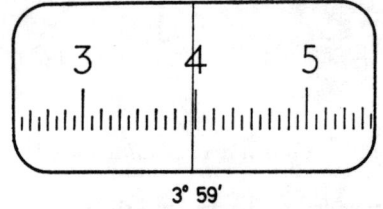

3° 59′

Optical scale. The microscope incorporates a finely sub-divided scale covering the same interval as one circle graduation.

Example: Wild T16 theodolite, circle graduated to one degree. The optical scale covers one degree, but it is sub-divided into single minutes. The reading is the main scale graduation cutting the optical scale *plus* the reading at the circle graduation on the optical scale. (Az = horizontal circle, V= vertical circle.)

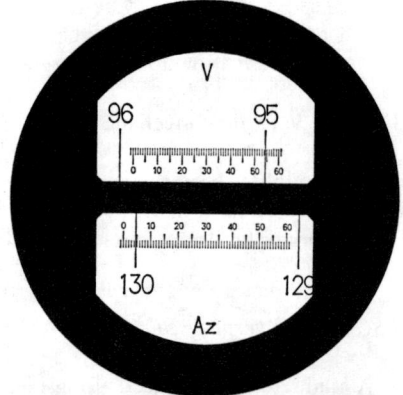

Vertical 95° 54.4′ Horizontal 130° 04.6′

Optical micrometer (single reading micrometer)
The microscope carries a fixed index line (or pair), and a circle graduation is brought into coincidence with the index line by the movement of a parallel-plate device actuated by a micrometer drum-head control. The amount of displacement of the circle image necessary to achieve coincidence is shown on an optical scale linked into the parallel plate mechanism, and the optical scale reading is added to the circle graduation reading.

Example: Wild T1-A theodolite horizontal circle. (Another coincidence setting needed to read vertical circle, using the same optical scale.)

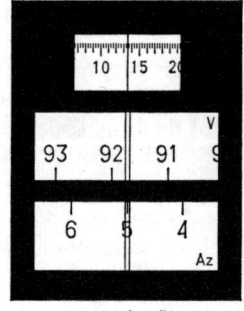

5° 13′ 35″

Coincidence optical micrometer (double reading)
The micrometer drum and parallel-plate arrangement is used to bring the circle graduation marks into coincidence not with a fixed index line but with the diametrically opposite circle graduations. The final reading is in effect, the mean of two diametrically opposed index marks, and it eliminates index error.

Example: Wild T2 theodolite. After coincidence, read 285°, count the divisions along to 105°, these are 5, hence add 50′. Read 1′ on lowest scale at index line, and read 54″·6 on the seconds scale.

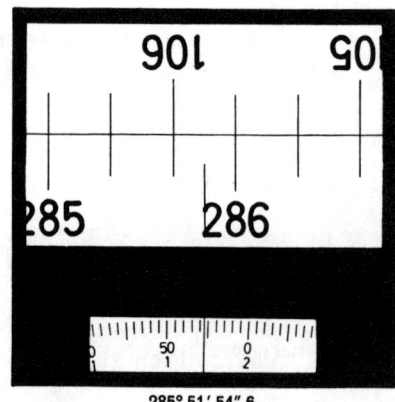

285° 51′ 54″.6

3·4 The surveyor's level

Three basic types today:

> The dumpy level
> The tilting level
> The automatic level.

All modern instruments use the internal focus telescope described earlier.

3·4·1 The dumpy level

Telescope is rigidly and permanently fixed to the vertical axis of rotation, and spirit level tube fixed to the telescope. Vertical axis supported and made level by a levelling head with three footscrews.

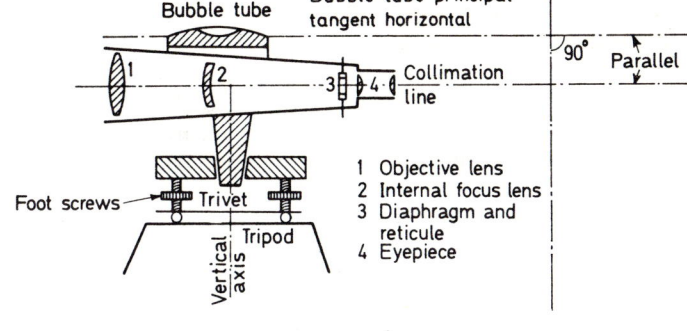

1 Objective lens
2 Internal focus lens
3 Diaphragm and reticule
4 Eyepiece

Essential operating conditions are:
(i) Spirit level tube principal tangent perpendicular to the vertical axis.
(ii) Collimation line parallel to the spirit-level tube principal tangent.

3·4·2 The tilting level

The telescope is hinged to a stage at the top of the vertical axis of rotation, and spirit-level tube fixed to the telescope. Telescope is tilted with respect to the vertical axis by the operation of the tilting screw and an opposed spring buffer. Vertical axis may be supported and levelled up by either a levelling-head and 3 footscrews or by a ball-and-socket mounting.

Essential operating conditions are: the spirit-level tube principal tangent must be parallel to the collimation line.

3·4·3 The automatic level

The telescope has a built-in automatic levelling compensator which, provided the telescope is levelled up within the working range of the compensator (usually about ± 20 minutes), automatic-

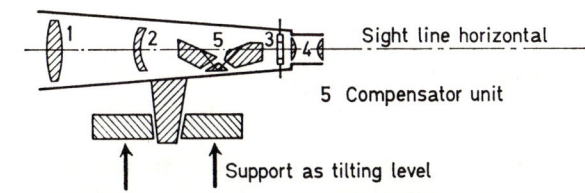

5 Compensator unit

Support as tilting level

ally ensures that the actual line of sight is horizontal, even if the telescope axis is not precisely horizontal.

The compensator unit generally consists of a system of fixed and suspended prism reflectors, usually three, and the triple reflection results in the telescope presenting an *erect* image at the eyepiece instead of the usual inverted image. The telescope is, in other respects, an internal focus type.

The vertical axis may be supported and levelled by three footscrews *or* a ball-and-socket mounting, with a circular reference spirit level.

Essential operating conditions are: The telescope and compensator should provide a horizontal sight line when the instrument is levelled up by reference to the circular (bull's eye) spirit level.

3.4.4 Plane parallel-plate micrometer

This is built into precise levels, but available as an attachment for other levels. Consists of a glass plate with plane parallel faces, placed in front of the telescope objective as in the figure.

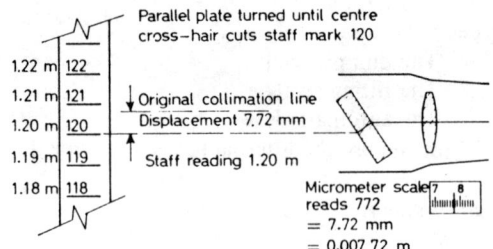

If the plate is tilted, the sight line is shifted up or down (down in the figure) but still parallel to its original direction. The plate is connected to a micrometer drum-head, and the actual shift in millimetres is either read off the drum edge or shown in an optical scale eyepiece alongside the telescope eyepiece. The micrometer is turned until the reticule line cuts a staff graduation, then the reading is staff graduation *plus* the micrometer reading. Normally used with a precise staff (*Invar*) marked at single centimetre intervals.

3.4.5 Classes of surveyor's level

Class 1 – Precise. Geodetic work or any precise levelling. May be tilting or automatic, but have large aperture, high magnification telescope and very high sensitivity bubble tube. Sight line typically levelled to ± 0·2 seconds of arc.

Class II – General purpose. Medium accuracy, for general engineering and non-geodetic work. May be tilting or automatic, few dumpys are made today.
Sight line levelled typically to ± 0·5 to 1 second of arc.

Class III – Builders. Low accuracy work such as ordinary building. May be tilting, automatic, or dumpy. Small, low-power telescope, typical aperture 20–30 mm and 10x to 20x magnification.

3·5 Permanent adjustments of the level

These are the adjustments necessary to keep the instrument in sound working order, they are carried out only when necessary, and they should not be confused with the temporary adjustments which are the operations required to set up the level ready for the day's work.

3·5·1 Dumpy level

Adjustment 1 – Spirit level tube principal tangent perpendicular to vertical axis.

Set-up the level on its tripod, with tripod head approximately level. Lay telescope parallel to footscrews 2 and 3, then follow the steps in the figure, from left to right, top row then bottom row. This covers the check and the actual adjustment. (Same method for theodolite plate bubble.)

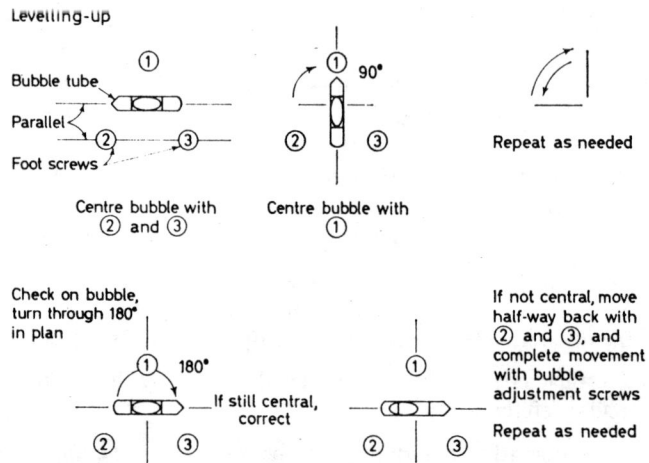

Adjustment 2 – Collimation line parallel to spirit-level tube principal tangent.

Check: Place pegs 1 and 2, at 50 to 100 m apart. Set up level midway between, level up carefully, read A1 and A2. Difference gives true level difference between pegs.
Set up as close as practicable to 2, on projection of line 1–2, level up carefully, read B1 and B2.
If instrument collimation line is horizontal, then $B2 = B1 + (A2 - A1)$.
If this not so, adjustment needed.

If collimation line at *B* is horizontal, then *B*2 should equal *B*1+(*A*2−*A*1)
Reading *B*1 should equal *B*2−(*A*2−*A*1)

To adjust: Shift the diaphragm, using its own adjusting screws, up or down as needed until the reading B1 is equal to $B2 - (A2 - A1)$.
Repeat the check and adjustment as needed.

Note: This is not strictly accurate, since the very small error in reading B2 has been ignored, but it has been demonstrated as adequate in practice, and it is suggested by some manufacturers.

3.5.2   *Tilting level*

One adjustment only: Spirit-level tube principal tangent parallel to the telescope collimation line.

Check: Carry out the two-peg test described immediately above for the dumpy.

To adjust: Tilt the whole telescope, using the tilting screw, until the centre cross-hair cuts the staff at 1 at the required reading B1, equal to $B2 - (A2 - A1)$. Finally, centre the spirit bubble using the adjusting screws fitted at one end of the bubble tube. Check and repeat as needed.

3.5.3   *Automatic level*

Adjustment 1 – Circular level bubble remain central when telescope turned about the vertical axis.

Check: Set up the instrument. Centre the circular bubble (footscrews or ball mounting, whichever fitted). Turn instrument through $180°$. If bubble remains central, no adjustment needed. Adjustment required if bubble moves off centre.

To adjust: If bubble off centre, move it halfway back to centre by means of the adjusting screws (three) provided in the underside of the bubble casing, then finally centre the bubble by means of the footscrews (or ball mounting, according to type).

Adjustment 2 – Collimation line horizontal when circular bubble centred.

Check: Check by the two-peg test.
To adjust: Move the reticule to correct reading using diaphragm screws.
Note: On some instruments the *compensator* and not the diaphragm should be adjusted – check with instrument handbook.

3.6 The theodolite

The figure shows the basic form of the earliest theodolite-type instruments. Modern instruments are refinements of this concept.

Pointing is by telescope, not open sights, vertical circle is a complete circle, the plumb-bob is replaced by more precise levelling and indexing arrangements. The supporting post is now a tripod, and many of the working parts are now fully enclosed.

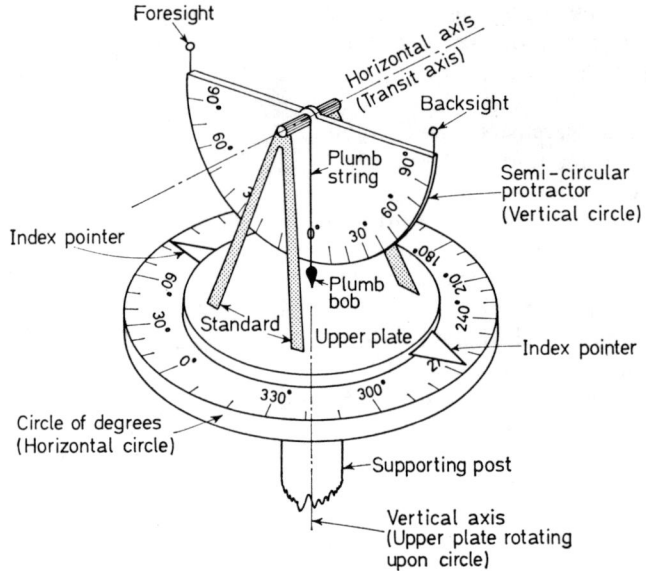

3.6.1 Vernier theodolite

Still widely used but being phased out of production. The figure shows the basic essentials of the instrument, with the omission of the circle clamps and slow-motion screws and the magnifiers for reading the vernier scales.

Standard internal focus telescope. Vernier indexes at both ends of a diameter of the upper plate — both should be read and the mean taken when measuring angles, to avoid errors due to eccentricity.

Normal vernier instrument is double-centre, both plates may be moved independently about the vertical axis of rotation, each plate having its own clamp and slow-motion screw. Other controls include clamp and slow-motion screw for vertical circle and a clip screw (altitude setting screw) which moves the vernier frame and the attached altitude bubble. Some models have centring motion provided *above* the footscrews, most are centred directly on the tripod head.

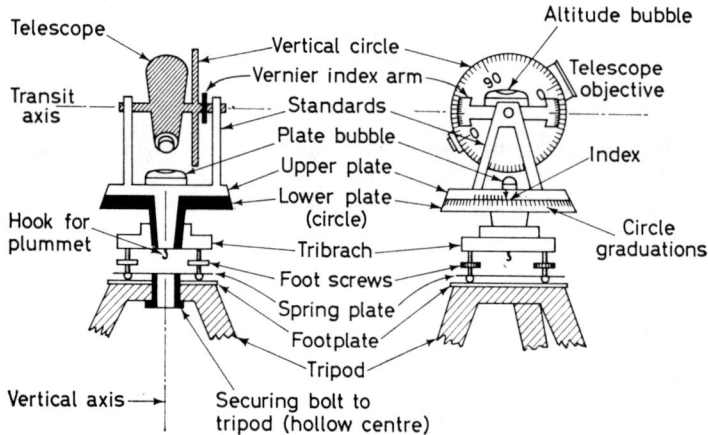

Setting up — Tripod placed over ground mark, instrument attached and roughly centred by plumb-bob. Instrument levelled by footscrews with plate bubble as in Section 3.5.1. Centring checked, adjusted if needed, levelling checked, etc.

The altitude bubble is centred by the clip screw before each vertical angle reading.

The upper plate and all parts it carries are often termed, together, the *alidade*.

Vernier theodolites may be classified by the diameter of the horizontal circle, e.g. '5 inch vernier', or by the least count of the vernier scales, typical descriptions being '30 second vernier theodolite' or '20 second'.

3·6·2   Optical reading theodolites

These are similar in principle to the vernier theodolite, but circles of optical glass and they are read by optical systems as described in Section 3·3·2. Lighter and smaller than vernier types, easier to read, and greater precision and accuracy obtainable with less labour.

An optical plummet often fitted allows better centring in windy weather. The tribrach generally detachable so instrument body can be replaced by target, subtense bar, etc. without altering centring. Automatic vertical circle index devices frequently fitted, eliminating the need for an altitude bubble and clip screw.

Double centre instruments have both upper and lower plates independently movable about the vertical axis, as the vernier types.

Repetition theodolites have the horizontal circle fitted between the alidade and the tribrach and capable of being clamped to either by a *repetition clamp.* One horizontal clamp and tangent screw control the motion of the alidade with respect to the tribrach. Particularly suited to *repetition measurements.*

Precision instruments usually have circle orienting motion — the lower plate cannot be swung but may be turned to any desired position by a drive screw, often termed the 'circle setting knob'.

3·6·3   Classes of theodolite

Class I – Precision. First and second order triangulation, geodetic astronomy, etc. Large aperture, high-magnification telescope, circles read by double-reading micrometer, direct to 0·5 seconds and by estimation to 0·1 seconds.

Class II – Universal (or 'Single second'). Second and lower order triangulation, traversing, high-accuracy engineering work. Telescopes only slightly below standards of class I, double reading micrometers reading to 1 second direct and to 0·5 or 0·1 second by estimation.

Class III – General purpose. Fourth order triangulation and general survey and construction work. Generally single reading micrometers, to about 20 seconds direct and 5 seconds by estimation. Sometimes optical scale, similar values.

Class IV – Builders. General low-accuracy work. Typical telescope 25—30 mm aperture, magnification 15x to 25x. Direct reading to 5 minutes, estimation to 1 minute or 30 seconds, optical scale or circle microscope.

3·7   Permanent adjustments of the theodolite

Principal adjustments:

     (a)   horizontal plate level perpendicular to vertical axis
     (b)   collimation line perpendicular to transit axis
     (c)   transit axis perpendicular to vertical axis
     (d)   vertical circle index adjustment.

The following summaries are appropriate to the common type of vernier theodolite and most optical theodolites equipped with an altitude bubble.

3.7.1 Horizontal plate level perpendicular to vertical axis

The check and adjustment are as detailed for the level in Section 3.5.1. If properly carried out, when the bubble is central it should remain so when the instrument is revolved in azimuth about the vertical axis.

3.7.2 Collimation line perpendicular to transit axis

Check — Set up. Clamp circle. Aim on a fine target at least 100 m distant. Clamp alidade, read circle, book as R1.

Unclamp alidade, transit, turn through 180°, sight target and clamp alidade. Read circle, book as R2.

If in adjustment, (R2 − 180°) = R1. If not, then Error = ½[(R2−180°)−R1].

To adjust — Calculate (R1+E), unclamp alidade, transit, point target and clamp. Using upper clamp and tangent screw, bring the circle reading to exactly (R1+E).

Move *diaphragm* laterally until central vertical cross-hair cuts the target. Check and repeat if needed.

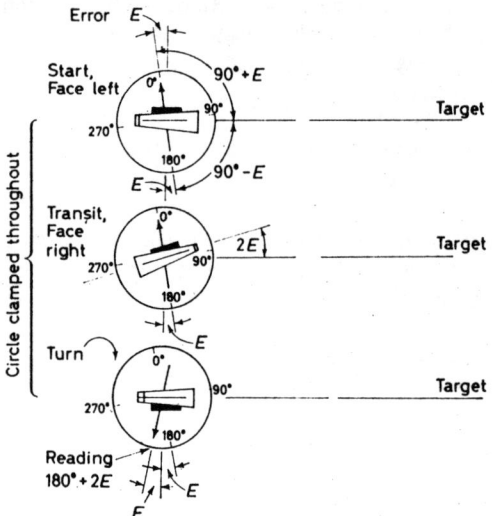

Note: For a *single reading instrument* (circle microscope, optical scale, single reading micrometer) the value of E must be found *again*, with circle turned through 180°, and the two values of E meaned before adjustment. This is needed to eliminate eccentricity errors. For vernier type, mean both vernier readings for R1, and again for R2.

3.7.3 Transit axis perpendicular to vertical axis

Check — Set up and level, about 15m from a tall building. With *face left* aim on fine target at 45° or more elevation. Finish with both plates clamped.

Depress telescope to sight on level staff at base of building, read S1.

Transit, turn through 180°, bisect high target again, plates clamped.
Depress telescope, read staff, book as S2. If in adjustment, S1 = S2.

To adjust — Calculate (S1 + S2)/2, mean reading. Bring centre vertical hair to mean reading, using upper clamp and tangent screw.

Elevate the telescope towards target. Tilt the transit axis by its adjusting screws until centre hair bisects target. Repeat check and adjustment as needed.

Note: No provision for transit axis adjustment on many modern instruments.

28

3.7.4 Vertical circle index adjustment

Check. — Set up and level. Point on a levelling staff 60–100 m distant. Centre altitude bubble by clip screw. Set vertical circle to $0°$ (or $90°$, according to type) such that telescope horizontal. Read staff at centre cross-hair, book as H1.

Transit, point staff again with changed face. Centre altitude bubble by clip screw. Set vertical circle to $0°$, $90°$, or $270°$ according to type (telescope horizontal). Read staff, book as H2. If in adjustment, H1 = H2.

To adjust — Calculate mean, (H1 + H2)/2. Set cross-hair to mean reading, using vertical circle clamp and tangent screw. Rotate clip screw until vertical circle is at zero position ($0°$, $90°$, $270°$, as appropriate). Centre altitude bubble by *bubble adjusting screws*. Repeat check and adjustment as needed.

3.8 Errors arising from theodolite maladjustment

Circle graduation errors: Small in modern instruments. To reduce effects, measure horizontal angles repeatedly on several parts of circle.

Circle eccentricity error: To reduce effects, mean both verniers or read on both faces for horizontal angles, mean both verniers on both faces for vertical angles. Cannot be eliminated in vertical angles with single reading instruments.

Index error: No effect on horizontal angles. Measure vertical angles on both faces.

Horizontal collimation error: To reduce effects, measure on both faces.

Plate level not perpendicular to vertical axis: No effect of importance in vertical angles. For horizontal angles, may have significant effect when objects viewed are at large angle of elevation or depression, but cannot be eliminated — either adjust plate level or apply correction to horizontal circle reading. To determine correction: Number bubble tube graduations, outward from centre of tube. On face left, note graduation numbers at left and right ends of bubble as L and R. On face right, note bubble end positions as L and R again (in each, left or right end as viewed from telescope eyepiece). Take mean of L readings and mean of R readings, then bubble shift = $(L_m - R_m)/2 = x$ graduations. Bubble dislevelment = x x bubble sensitivity seconds.
Correction to mean horizontal circle reading = x x sensitivity x tan(vert. angle).
(Always subtract R_m from L_m and observe correct sign).

Transit axis error: Eliminate by measuring on both faces. If axis tilt actually due to plate bubble error, proceed as for plate level error.

3.9 Angle measurement terms

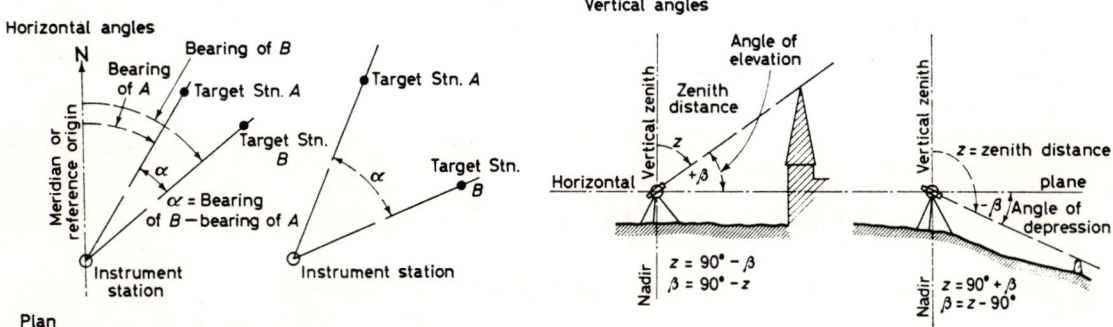

3·10 Angle measurement methods

Vertical angles: Measure on both faces, take the mean. This eliminates all maladjustment errors except circle graduation error. To reduce reading errors, take several sets of observations.

Horizontal angles: Measure on both faces, take the mean. This eliminates all maladjustment errors except circle graduation errors and plate level error. The latter is not often significant, and circle graduation and reading errors are minimised by taking several sets of observations on different zeros. The methods used to do this are (a) simple reversal, or (b) the method of repetition.

3·10·1 Simple reversal

The figure shows simple reversal — angle measured twice on the same zero, once face left, once face right. Results meaned.

If n sets of observations to be made of the angle, change zero for each new set by $180/n$ degrees.

The figure is for single-reading instrument, for vernier type there will be two index readings to be meaned for each pointing.

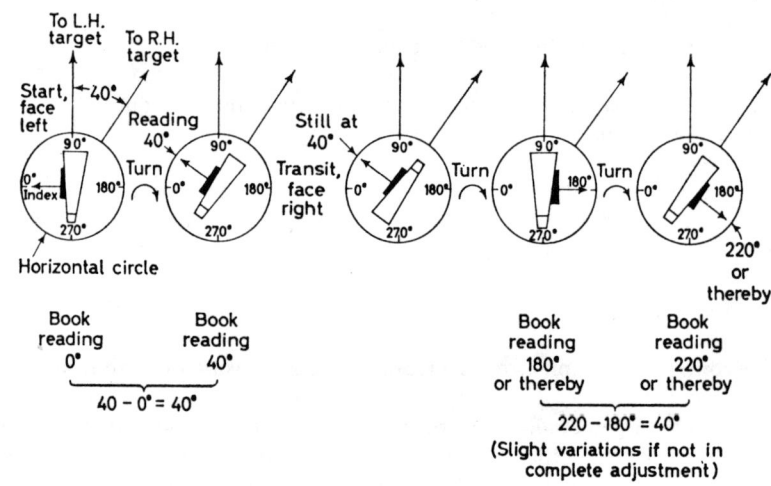

3·10·2 Method of repetition

The figure shows the angle turned off twice on the circle, final reading divided by 2.

For high accuracy the angle turned off n times with face left, telescope transited, then angle turned off n times with face right and swinging left.

This method best suited to small angles, such as parallactic angles in subtense measurement.

4 Levelling

Levelling – The determination of the relative heights (or differences in height) of points on the surface of the earth.

Level line – A line lying throughout on one level surface and normal to the direction of gravity at all points.

Level surface – A surface which is normal to the direction of gravity (as defined by a plumb-bob) at all points on the surface.

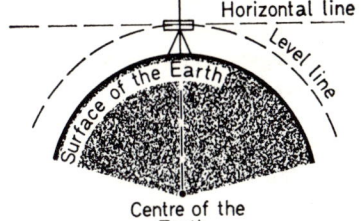

Datum surface or line – Any level surface or line, of known height, from which heights or levels are measured.

Horizontal line – A straight line through a point, tangential to the level line through the same point, and normal to the direction of gravity at the point.

Reduced level of a point – Height of the point above the particular datum in use.

Benchmark – Fixed point of known height, from which other levels may be established.

Temporary benchmark – Temporary B.M. set up by a surveyor for his own use on a particular job.

Backsight – The first sight taken after setting up a levelling instrument.

Foresight – The last sight taken before moving the instrument.

Intermediate sight – Any levelling sight which is neither a backsight nor a foresight.

Changepoint – Stable point on which the level-staff is held while the level is moved from one set-up point to another. (Cardinal rule: Never move level and staff at the same time while levelling is in progress.)

Spirit levelling – Levelling operations carried out with the surveyors' level.

4·1 Levelling procedure and booking

Figure shows series levelling, a line of levels starting at a B.M. of known height.

Several instrument set-ups required, a changepoint between each pair. Instrument need not be set up on the line of levels, but placed for best visibility.

One standard method of booking the field observations, but two methods of reducing these to obtain the heights of the different points.

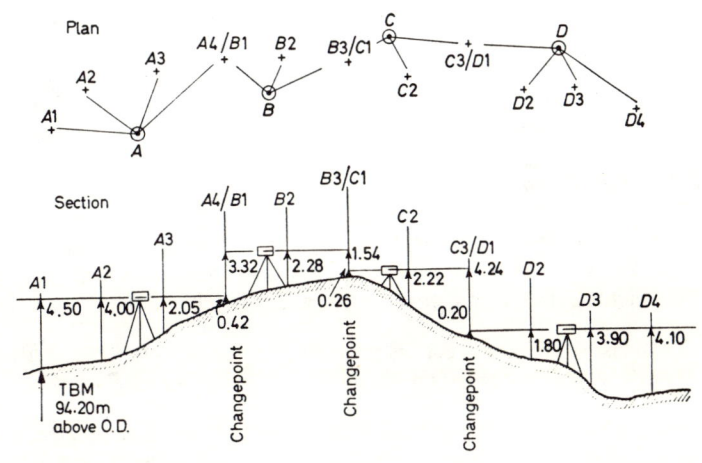

31

Important rule in booking — Enter all readings and detail regarding one ground point *on the same line of the level book.*

4.1.1 Rise and fall reduction method

Enter 'rises' and 'falls'. A rise is the vertical movement of the staff in going from one point to a *higher* point; a fall is the vertical movement in going to a *lower* point.

Total backsight column, total the foresight column, take the difference. Total rises, total falls, take difference, should agree with that of back-and-foresight columns. If not, check all the calculations.

Rise and Fall

Back	Inter	Fore	Rise	Fall	Red. Level	Dist.	Remarks
4.50					94.20		TBM 1 A 1
	4.00		0.50		94.70		£ road A 2
	2.05		1.95		96.65		" " A 3
3.32		0.42	1.63		98.28		CP " " A4/B1
	2.28		1.04		99.32		" " B2
0.26		1.54	0.74		100.06		CP " " B3/C1
	2.22			1.96	98.10		" " C2
0.20		4.24		2.02	96.08		CP " " C3/D1
	1.80			1.60	94.48		" " D2
	3.90			2.10	92.38		" " D3
		4.10		0.20	92.18		TBM 2 at gate post D4
8.28		10.30	5.86	7.88	94.20		
		8.28		5.86	2.02		
		2.02		2.02			Check O.K.

Calculate reduced levels, check that difference between the last and the first agrees with the other differences. If agrees, the arithmetic, but not necessarily the levelling, is correct.

4.1.2 Collimation height method

Calculate the height of collimation line on first set-up. Deduce levels of all other points observed from that set-up.

Calculate collimation height of next set-up, deduce reduced levels.

Proceed in the same method throughout.

Check is carried out as shown, if all intermediates are to be checked.

The rise and fall method provides an easier and more immediate check on the arithmetic and is therefore the preferred method.

Collimation Height

Back	Inter	Fore	Coll. Height	Red. Level	Dist.	Remarks
4.50			98.70	94.20		TBM 1
	4.00			94.70		£ road
	2.05			96.65		" "
3.32		0.42	101.60	98.28		CP " "
	2.28			99.32		" "
0.26		1.54	100.32	100.06		CP " "
	2.22			98.10		" "
0.20		4.24	96.28	96.08		CP " "
	1.80			94.48		" "
	3.90			92.38		" "
		4.10		92.18		TBM 2 at gate post
8.28		10.30		94.20		
		8.28		2.02		Sum of each collimation height
		2.02				X no. of reduced levels from it = 988.78

Sum of all intermediate sights = 18.25
"   "   " foresights   = 10.30
"   "   " RL's exept first = 962.23   Check OK
= 988.78

4.1.3 Page by page check

Check each page in the level book separately, before proceeding to the next page. Start every page with a backsight, finish every page with a foresight reading. Check that each page has the same number of backsights as foresights, or the page cannot be checked.

If the last reading on a page is actually an *intermediate sight*, write it in the *foresight column*, and enter it again as the first entry on the next page in the *backsight column.*

Always leave at least four lines clear at the bottom of each page, to leave adequate space for the check calculations. Do not erase errors — cross out and enter revised values above.

4·2   Curvature and refraction

Curvature of the earth and refraction of light by the earth's atmosphere both affect level readings and vertical angle measurements. Their effects may be neglected in short range work but must be allowed for in long sights and precise work.

4·2·1   Curvature

Effect on level staff readings — AB represents horizontal sight line at A. Level staff placed at B_1. Staff reading in error by the amount BB_1.

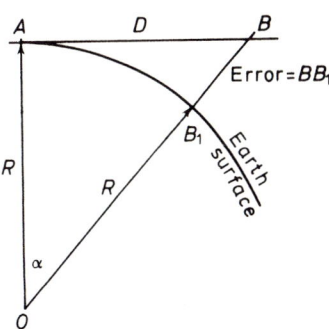

$(R + BB_1)^2 = R^2 + D^2$, and $(BB_1)^2$ is negligible,

∴ $BB_1 \simeq D^2/2R$

Error eliminated in levelling if backsight distance and foresight distance equal.

On a *line of levels,* errors cancel if total backsight distances equal total foresight distances.

Taking earth radius as 6370 km, curvature error $= -0.0785\,D^2$ *metres*, where D is stated in kilometres.

Effect on vertical angles — Vertical angle measured with respect to the horizontal line AB, angle θ. Should be referred to line AB_1 which approximates to the earth surface between A and the ground at B_1 below the target.

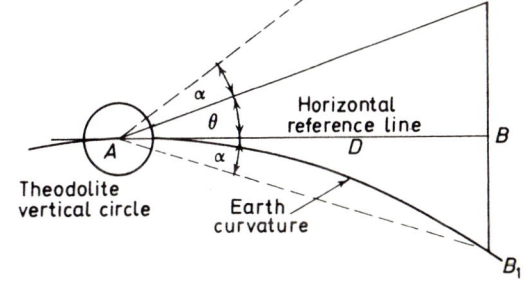

$BB_1 = D^2/2R$ = Linear amount of earth curvature.

Curvature angle $\alpha = D/2R$,

(since α is small, $\tan \alpha = \alpha$ radians) and

$\alpha'' = 206265\,D/2R$ seconds.

This may be expressed as $\alpha'' = PD$, where P is the constant $206265/2R$.

The observed angle θ may be positive or negative, according to whether it is an angle of elevation or depression. The correction α'' is always positive.

4·2·2   Refraction

When considering the sight line from one station to another, the *refraction angle, r,* is the angular distance between the observed and the true line of sight. The *coefficient of refraction, k,* is the ratio of the refraction angle to the angle subtended by the two stations at the centre of the earth.

The value of k varies with the time of day, the rate of change of temperature with altitude, the length of the line, the type of terrain, and the climate. An average value is 0·07.

Effect on levelling – AB represents horizontal sight line at A. Level staff placed at B_1. The *actual sight line* is AB_2, the light being refracted through the angle r, and depressed by distance $e = BB_2$.

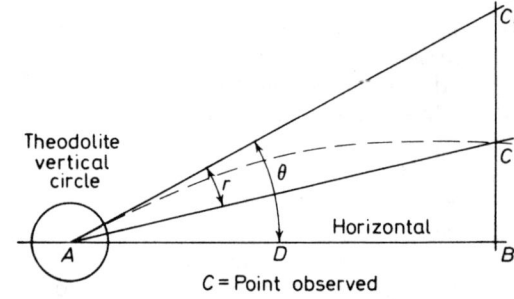

Angle $r = k$ x angle α,

and since α is small, $\tan \alpha = \alpha$ radians,

$$\therefore \quad \text{angle } r = k \text{ x } \tan \alpha$$
$$= k \text{ x } D/R,$$
$$\text{also } e/d = \tan r = \text{angle } r$$
$$\therefore \quad e/D \quad = kD/R$$
$$\therefore \quad e \quad = kD^2/R.$$

If levelling carried out from early morning to noon, the value of k and therefore the amount of refraction, will steadily diminish, and error in foresight reading due to refraction will always be less than in the backsight reading. To compensate, use two level staves and always read the same staff first at each set up.

If levelling up a regular incline, the foresight (nearest to ground) will always be refracted less than the backsight, hence uphill end of the line will be too low.

General rule to minimise refraction effect is to level a line twice, under the same conditions for both the forward and backward runs.

Effect on vertical angles. – Vertical angle measured to point C, but the telescope is actually pointed in the direction to C_1, recording the angle θ. Light path curved down to C due to refraction.

Refraction angle $r = k$ x D/R
$$\therefore \quad r'' \quad = kD \text{ x } 206265/R$$
$$= 2\, kPD,$$

Where P is the constant $206265/2R$ as expressed in Section 4.2.1.

The observed angle θ may be positive or negative, according to whether it is an angle of elevation or depression, but the correction r'' is always negative.

4.2.3   Combined curvature and refraction

Effect on levelling – AB represents horizontal line at A, actual line of sight being AB_2.

$$\text{Curvature correction} = D^2/2R.$$
$$\text{Refraction correction} = kD^2/R.$$
$$\text{Combined correction} = D^2/2R - kD^2/R$$
$$= D^2(1 - 2k)/2R$$

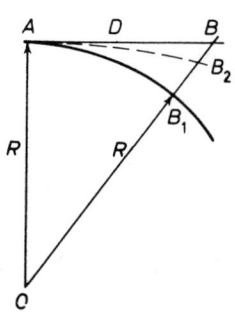

Taking $k = 0.07$, $R = 6370$ km, combined correction = $0.0673\, D^2$ *metres,* where D is the sight distance in kilometres.

Easier version to remember, error $\approx - D^2/15$ metres, D being stated in kilometres.

Effect on vertical angles. — The two corrections for curvature and refraction may be combined directly.

Corrected angle   =   observed angle + curvature corr. − refraction corr.

or ϕ = $\theta + PD - 2\,kPD$

   = $\theta + PD(1 - 2k)$.

Where P is the constant $206265/2R$ of Section 4·2·2 and 3.

If $k = 0.07$, $R = 6370$ km, then the correction amounts to $+ 13.91$ seconds of arc per kilometre of distance, always positive.

4·2·4   *Intervisibility problems*

Typical problems of the form illustrated — sighting from A, height at point B needed to ensure visibility, or the height of signal needed at C.

Neglecting curvature and refraction, use tangent or sine of θ.

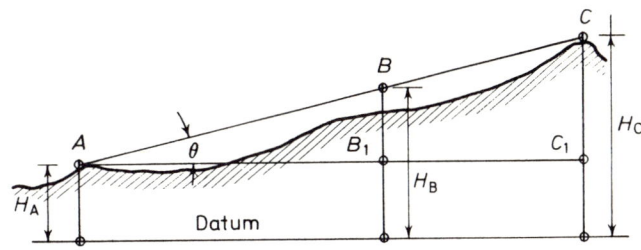

$$\tan \theta = (H_C - H_A)/AC_1 = (H_C - H_B)/B_1C_1 = (H_B - H_A)/AB_1$$

$$\sin \theta = (H_C - H_A)/AC = (H_C - H_B)/BC = (H_B - H_A)/AB.$$

In practice, the combined effect of curvature and refraction is to curve the datum ABC into the shape A_1BC_1, where
$AA_1 = 0.067 \, (AB)^2$ metres,
$CC_1 = 0.067 \, (BC)^2$ metres,
both AB and BC being measured in kilometres.

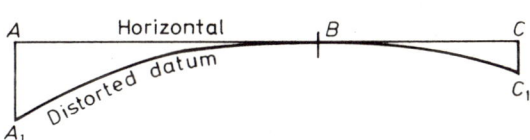

Referring to the first figure, the net effect of curvature and refraction is to *lower* the datum at A and C with respect to B. If therefore the heights of A and C are *reduced* by $0.067 \, (AB)^2$ and $0.067(BC)^2$ metres respectively (AB and BC being stated in kilometres), the tangent or sine equations may be applied.

4·3   Sources of error in levelling

Instrumental errors — Faulty permanent adjustments.
To reduce effects, check permanent adjustments, equalise back and foresight lengths.

Instrument manipulation errors — Parallax, non-central bubble, tripod movement.
To reduce effects, check all operations, ensure set-up on stable ground.

Staff manipulation errors — Non-vertical staff, sinking in soft ground, not fully extended (telescopic staff).
To reduce effects, swing staff or use staff bubble, use change plate, check catches.

Reading and booking errors — Check and practice.

Natural causes — Differential expansion of instrument, heat shimmer, wind, curvature and refraction.
To reduce effects, use umbrella, avoid grazing rays, shield instrument if necessary, equalize backsight and foresight lengths.

4·3·1   General rules to minimise error

(a)   To detect gross errors, run a line of levels between two points of known height, alternatively run twice over the same line in opposite directions but under similar climatic conditions.
(b)   Always check that the staff is vertical when reading, either by swinging or using staff bubble.
(c)   Always check that backsight and foresight lengths are equal for any one instrument set-up.

4·3·2   Accuracy in spirit levelling

Closing error estimate is quoted in the form $E = \pm c \sqrt{K}$ mm, where c is a constant fixed by the type of work and conditions and K is the distance levelled in kilometres.

Typical allowable closing error values are $\pm 20 \sqrt{K}$ mm, for ordinary work and rough ground, and $\pm 10 \sqrt{K}$ mm for careful work on reasonably flat ground.

4·4   Levelling over wide gaps

If back and foresight lengths cannot be equalised, reciprocal observations are necessary to eliminate collimation, curvature and refraction errors.

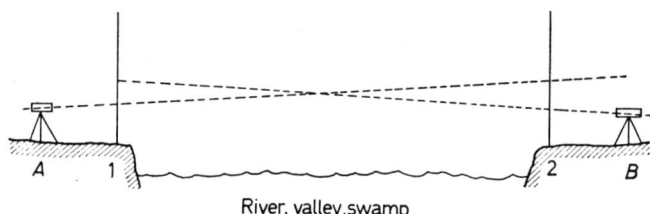

River, valley, swamp

4·4·1   Reciprocal levelling

Referring to the figure above, readings from A on staves at 1 and 2 give height difference between points 1 and 2. With instrument moved to B, readings on 1 and 2 give another value for the height difference between points 1 and 2. The *mean* of the two values is free from collimation and curvature error, and also free from refraction error provided the refraction coefficient does not change during the operations. Staff target necessary on long sights.

4·4·2   Gradienter levelling

Using tilting level with gradienter screw marked in tangent parts per thousand, gradienter reading t represents an angle whose tangent is $t/1000$.

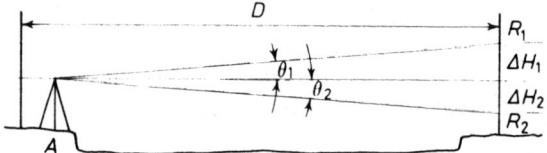

Level set up at A, staff on distant point, and target placed on staff at reading R_1. Telescope sighted on target, gradienter read, t_1.
Since $\theta \simeq \tan \theta$,

$$\tan \theta_1 = (t_1 - t_0)/1000, \text{ and } \Delta H_1 = D(t_1 - t_0)/1000,$$

where t_o is gradienter reading with bubble centred.
Taking a second reading, with t_2 negative,

$$\tan \theta_2 = (t_2 - t_0)/1000, \text{ and } \Delta H_2 = D(t_2 - t_0)/1000$$

The two values for the horizontal sight reading are $(R_1 - \Delta H_1)$ and $(R_2 + \Delta H_2)$, the mean value $\frac{1}{2}[R_1 + R_2 - (t_1 + t_2 - 2t_0)D/1000]$ being used as a normal staff reading.

Procedure repeated from the opposite side, again all errors but refraction are eliminated.

4·4·3   Simultaneous reciprocal observations

If one instrument used, refraction not eliminated. To eliminate refraction, use two instruments, one each side of gap, read simultaneously. To allow for instrument error, change over and repeat again. As many sets of observations taken as necessary to achieve the required accuracy.

4·5   Applications of spirit levelling

(a) Section levelling for roads, sewers, railways, etc.
(b) Area levelling for contours, spot heights, earthwork volumes, etc.
(c) Setting-out levels for construction works.

4·5·1   Section levelling

Longitudinal section – Ground profile along longitudinal line of any proposed construction. May include actual ground levels, formation levels, invert levels, gradients, cut and fill data.

Example shows common method of plotting, the vertical scale normally much larger than the horizontal.

Cross-section – Ground profile at right angles to the longitudinal line of the construction. Generally same scales both vertically and horizontally.

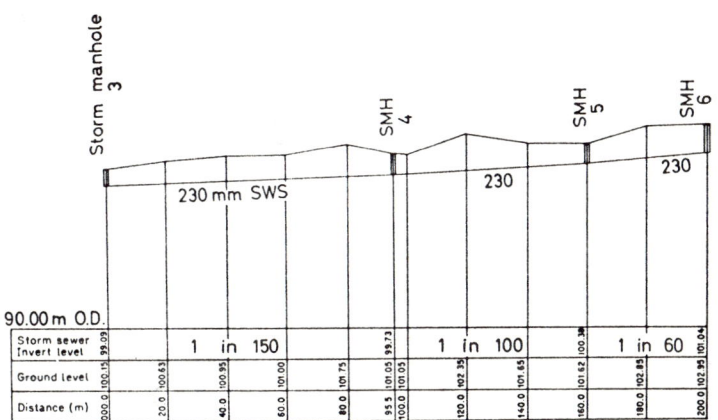

Longitudinal Section: Storm sewer 'A', CHGE. 0 to 2 + 00

Scales: Horizontal 1 : 500 (original plotting scales)
         Vertical 1 : 100

(All measurements in metres)

4·5·2   Area levelling

Contour line – Line on drawing representing an imaginary line on the ground connecting all adjacent points of equal height.

Vertical interval – The difference in height between successive contour lines.

Horizontal equivalent – The shortest horizontal distance between two contour lines.

Direct contouring – Location of contour lines directly upon the surface of the ground. Must be followed by plan survey to locate the points on the plan.

Indirect contouring – Determination of the heights of a number of points on the ground, these being marked on the plan and the contours interpolated between them. Point location may be by grid established with ranging rods and tape or chain, by radiating lines from a central instrument station, by sections of levels, or by radiation using stadia measurement. (Tilting level with horizontal circle, bearings noted and distances obtained from stadia lines.)

4·6   Trigonometrical levelling

Determination of height differences by the measurement of vertical angle from one station to another, the distance between the stations in plan being known.

4·6·1   Basic principle

(Neglecting curvature and refraction)

Theodolite over station A, target over station B, required quantity is
$$\Delta H_{AB} = H_B - H_A.$$
Horizontal distance between stations is D, vertical angle at A is θ_A.

$$H_A + i_A + D \tan \theta_A = H_B + m_B$$

$$\therefore \quad \Delta H_{AB} \stackrel{\cdot}{=} D \tan \theta_A + i_A - m_B$$

Sign conventions: Heights above a station are positive, below it negative. Angle of elevation positive, depression negative.

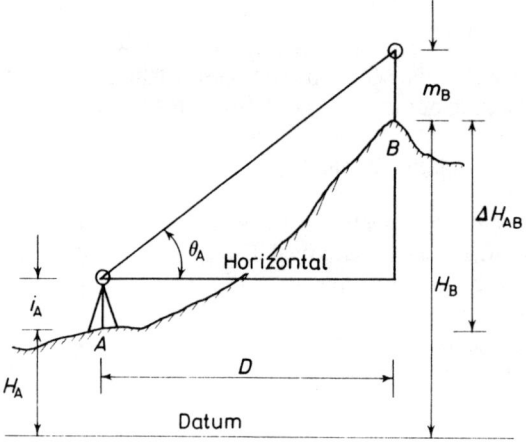

4·6·2   Curvature and refraction

The vertical angle must be corrected for curvature and refraction, then

$$\Delta H_{AB} = D \tan \phi_A + i_A - m_B.$$

where
$$\phi_A = \theta_A + PD(1-2k), \text{ and } P \text{ is the constant } 206265/2R.$$

4·6·3   Alternative methods

Single observations – Observations made from one end of the line, such as A in the figure above.

$$\Delta H_{AB} = D \tan \phi_A + i_A - m_B,$$

with ϕ_A being the corrected vertical angle at A.
This relies on an accurate determination of k, which is unlikely.
A check on *gross error* would be provided if a further measurement is made from end B.

Simultaneous reciprocal observations – Simultaneous observations made from A towards B and from B towards A. If the observations are truly simultaneous then the mean of the two results is free from error due to refraction.
Multiple sets of observations made according to accuracy demanded.

Approximate methods – For short lines, adequate to use the method of section 4·6·1 and add on the linear correction for curvature and refraction, then

$$\Delta H_{AB} = 10^3 D \tan \theta_A + 0 \cdot 067\, D^2 + i_A - m_B \text{ metres}$$

where D is the distance in kilometres.

For very short lines of only a few hundred metres, the second term may be neglected and

$$\Delta H_{AB} = d \tan \theta_A + i_A - m_B \text{ metres}$$

where d is the distance in metres.

38

5 Plane Table Survey

Survey method in which the angles or directions between survey stations or detail points are drawn directly onto a portable drawing table in the field. May be combined with linear distance measurement (direct or optical). Formerly widely used for topographical map detail, now widely replaced by air photo survey for this purpose.

5·1 Equipment

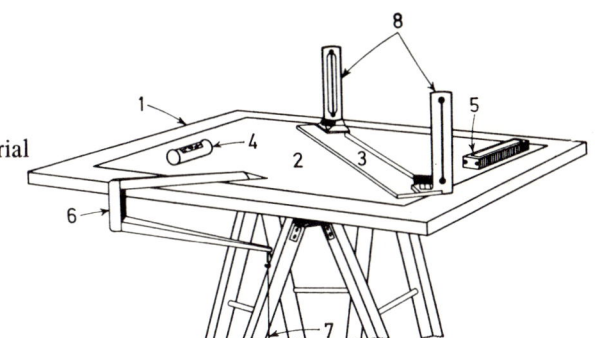

Basic equipment shown in the figure.

1. the plane table
2. paper or other draughting material
3. simple alidade
4. spirit level
5. box compass
6. plumbing fork
7. plumb string
8. sighting vanes on alidade.

Telescopic alidade frequently used today, similar to theodolite alidade rigidly fixed to sighting rule and complete with vertical circle. The Beaman Stadia Arc (see Section 8·1·2) is fitted to some telescopic alidades. The best form of telescopic alidade is the Self-Reducing Tacheometric Alidade as outlined in the same section.

5·2 Survey methods

These are:

    (a)   radiation   (c)   intersection/triangulation
    (b)   traversing   (d)   resection.

5·2·1   Radiation

Figure shows table set up at central station, capital letters represent ground points. Alidade pointed on station, line drawn along straight edge, distance to station measured by tape or band and scale distance plotted on board. Used to fix stations or close detail.

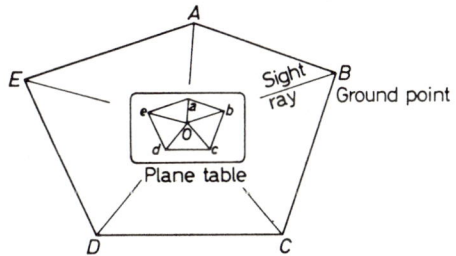

5·2·2   Traversing

The table is set up at each station in turn, rays being drawn to the forward and back stations and the distances to these being measured and set out by scale. See left-hand figure below. The right-hand figure shows graphic adjustment of plane table traverse. Closing error dd_1, other stations proportionally adjusted in parallel direction.

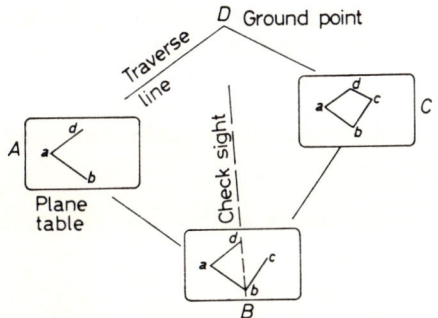

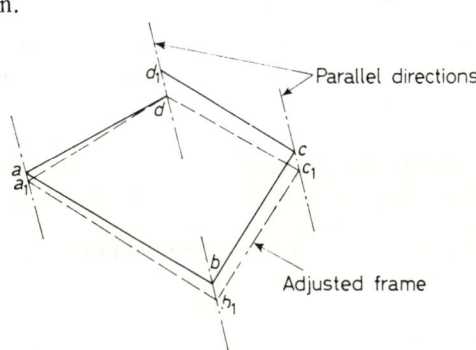

5.2.3 Intersection/Triangulation

Point located by intersection of two (or more) rays from different plane table stations. At least two ground stations required and the distance between them (base-line) but distance to observed points need not be measured.

At least three rays required to fix a further survey station.

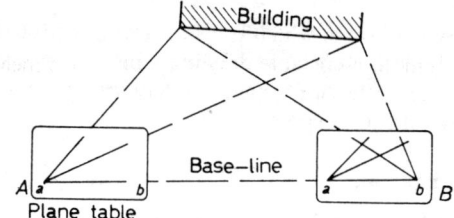

5.2.4 Resection

Used to fix the position of the station actually occupied by the plane table.

The figure shows *Three-point resection*, and a, b, and c are plotted points on the plan, all visible.

Select trial point O on paper to represent the occupied station point.

Using alidade, draw light line from O to *a*. Rotate table until this line is sighted at the distant point A. Put pencil on *b*, lay alidade against pencil, sight on B, and draw line against alidade to intersect the first line. Repeat with point *c* to produce a third line.

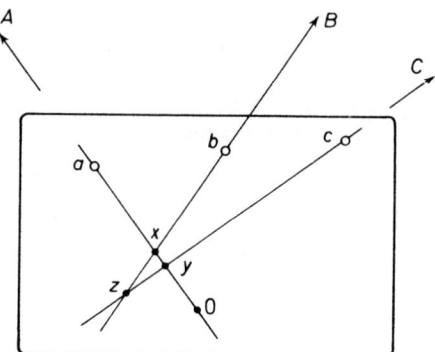

The three lines now constitute a *triangle of error*, with x, y, and z being the intersection points of the lines. Scale the distances Oa, Ob, and Oc, let these be termed d_1, d_2, and d_3.

> *Three rules for location of the station on the plan by trial and error*
> (a) If O is within the triangle of error then the correct station point lies within the triangle of error. If O lies outside the triangle then the correct position is outside the triangle.
> (b) If O lies outside the triangle, then the correct point lies either to the left or to the right of all the rays, looking towards the fixed points.
> (c) The correct station point is located such that the distance of the point from the three rays is proportional to the distances d_1, d_2, and d_3.

Having decided on the point, mark it as O_1. Lay the alidade along $O_1 a$ and re-orient the table. Finally check by sighting $O_1 b$ and $O_1 c$.

Note: No solution possible if table lies on circle through A, B, and C. *Two-point resection* is an alternative method based on two fixed points, but it requires the use of an auxiliary point.

5.3 Control for plane table survey

In the traditional topographical mapping by plane table the plane table stations were fixed by triangulation, the plane table merely being used for detail fixation. Extra table stations may be located by graphical triangulation or resection as needed.

An existing plan may be placed on the plane table and extra (or altered) detail filled in by plane table methods.

An independent plane table survey without external control must be limited in extent since error builds up rapidly with graphical methods.

5·4   Contouring

If the heights of the plane table stations are known, the surveyor may locate a large number of spot heights and sketch the contours directly in the field. An alternative is to contour direct.

Spot heights may be fixed by Indian Clinometer, which gives the tangent of the vertical angle observed, or by clinometer (Abney Level) or any of the forms of telescopic alidade.

An alternative method for nearby points is to use the surveyor's level for height and distance (in conjunction with a level staff), direction to the point being fixed in the usual way.

5·5   Advantages and disadvantages of plane table survey

Advantages

    (a)   all plotting in the field, less chance of omitted detail
    (b)   field notes kept to minimum, booking errors eliminated
    (c)   office work reduced to minimum, information immediately available
    (d)   the method is rapid and easy.

Disadvantages

    (a)   impossible in wet and windy weather
    (b)   equipment cumbersome, small parts easily lost
    (c)   lack of field notes sometimes inconvenient, e.g. for area/volume calculations
    (d)   impractical in heavy woods or dense bush.

6 Traverse Survey

Survey method in which the framework is formed by a series of straight lines, the length of each line and the angle between successive lines being measured.

Closed traverse — One commencing on a fixed survey station and terminating on another fixed station. (Also termed a *controlled traverse.*) If the traverse closes back onto the initial station it may be described as a *closed-loop traverse.*

Open traverse — An uncontrolled traverse which does not terminate on fixed survey points at both ends.

6·1 Traverse angle measurement

Methods used for angle measurement:
  (a) magnetic bearings — compass, or theodolite with compass, either 'fixed' or 'free' needle
  (b) included angles — by theodolite
  (c) deflection angles — by theodolite
  (d) direct bearings — by theodolite.

6·1·1 Bearings

The directions of the lines of a traverse are normally stated with respect to the reference direction (meridian) chosen for the whole traverse. A common reference direction is essential for any traverse which is to be co-ordinated.
Whole-circle bearings are measured clockwise from north, $0°$, through east, south and west to $360°$ (north) again.
Quadrantal or *reduced bearings* are measured from north or south, either in an easterly or a westerly direction, with a maximum value of $90°$.

Bearings measured in the field are normally stated as *whole-circle bearings,* and for the computation of co-ordinates these are converted to *reduced bearings.*

Where a line is specified by stating the stations at each end of the line, the bearing of the line is assumed to be measured at the first station and looking towards the second station. This may be more correctly termed the *forward bearing* of the line. The back bearing of a line is equal to its forward bearing $\pm 180°$ in the whole-circle system.

6·1·2 Reference meridian

The *reference meridian* of a traverse is the reference direction chosen for all the line bearings of the traverse.
Reference meridian chosen may be:
  (a) true meridian
  (b) magnetic meridian
  (c) grid meridian
  (d) any arbitrary meridian.

True meridian
The true meridian at a point is the north-south direction defined by a plane passing through the north and south astronomical poles and the given point. The true meridian through a point can be determined by astronomical observations, and it does not vary.
Since meridians converge from the equator to the poles, the true meridians through several survey stations are not parallel and the meridian through one particular station must be chosen as the reference meridian for the whole survey.

Magnetic meridian.
The magnetic meridian at a point is the direction which is indicated there by a properly balanced freely floating magnetic needle, assuming the needle is free from local disturbance.

Magnetic declination at a point is the horizontal angle between the true and magnetic meridians through the point. Declination at a point is not constant but is subject to regular and irregular variations.
Isogon is a line (imaginary) passing through points at which magnetic declinations are equal at a given time.
Agonic line is an isogon through places at which magnetic declination is zero.

Regular variations in declination

Secular variation. A slow continuous swing with a period of several centuries.

Diurnal variation. Variation from the mean value during any one day. The mean value occurs in the U.K. about 10 a.m. and again about 6 to 7 p.m. Needle swings eastwards at night, maximum about 8 a.m., then westwards with a maximum about 1 p.m.

Irregular variation – is caused by magnetic storms.

Grid meridian.
National maps are normally based on a rectangular grid, the centre north-south grid line being the true meridian through the centre of the mapped area. (U.K. grid is based on the true meridian $2°$ W of Greenwich.) *Grid north* at a point is a line through the point and parallel to the central meridian of the grid. *Grid bearing* is a bearing reckoned from the direction of *grid north*.

The difference between true and grid bearing at a point is the *convergence* at the point. In minor work it may be taken that the convergence of meridians at a point is approximately equal to the sine of the latitude of the point multiplied by the difference of longitude between the point and the central meridian.

Arbitrary meridian.
Any convenient direction may be assumed as the reference meridian for a traverse. An arbitrary meridian may be converted to a true meridian if desired.

6·1·3   Included angle measurement

The included angle at a traverse station may be either of the two horizontal angles between the two survey lines meeting at the traverse station. As a general rule, it is best to measure the included angle *clockwise* from the back station to the forward station.

The included angles may be measured once only, or once face left and once face right, according to the accuracy required of the work.

Included angle measurement is preferred for long traverses or higher accuracy.

6·1·4   Deflection angle measurement

The deflection angle at a traverse station is the angle between the back line produced and the forward line. It is therefore equal to the difference between the included angle and $180°$.

To measure the deflection angle at a station, sight theodolite on back station with all clamps tight, transit the telescope and sight on forward station using upper clamp and tangent screw. The difference between the circle readings when pointed on the back station and the forward station gives the deflection angle. A note *must* be made to indicate whether the deflection is to the right or left of the back line produced.

Deflection angles are used in some engineering survey work, but in general it is best to use included angle measurement.

6·1·5   Direct bearing measurement

Direct bearing measurement in traversing means that the theodolite is manipulated in such a way that when it is set over a traverse station and the telescope is pointed in a particular direction then the horizontal circle reading will give the bearing of that direction as measured from the reference meridian of the survey.

Three methods of direct bearing measurement:
> (a)   direct method with transiting
> (b)   direct method without transiting
> (c)   back bearing method

In each case the observations may be made on one face or, for higher accuracy, on both faces.

Direct bearing methods are generally used on short or lower-order traverses.

Direct method with transiting
The following notes assume the theodolite is set up at station A, and oriented with respect to the reference meridian, and later moved to stations B, C, etc., in turn. Single face observation only.

At station A:
(1) Telescope pointed on B, bearing AB recorded. Other station observation carried out as needed.
(2) Telescope transited, alidade rotated until circle reads bearing AB ± 180°.
Check that station B is bisected by central vertical cross-hair of telescope reticule.
(3) Move theodolite to station B.

At station B:
(1) Set up and level.
(2) With the same face as (1) above, and alidade still clamped, point A using lower clamp and tangent screw, thus the circle reads bearing AB ± 180°, i.e. bearing BA.

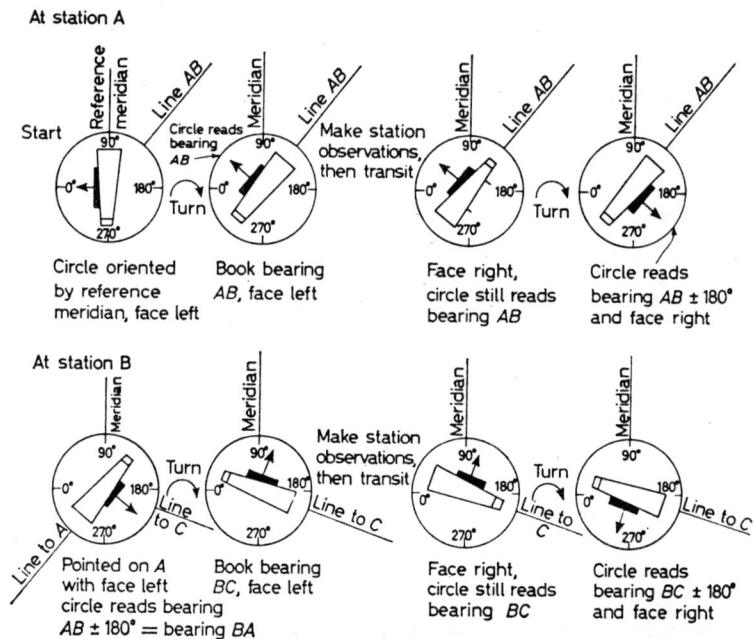

At station A

Start — Circle oriented by reference meridian, face left
Book bearing AB, face left
Make station observations, then transit — Face right, circle still reads bearing AB
Circle reads bearing AB ± 180° and face right

At station B

Pointed on A with face left circle reads bearing AB ± 180° = bearing BA
Book bearing BC, face left
Make station observations, then transit — Face right, circle still reads bearing BC
Circle reads bearing BC ± 180° and face right

(3) Turn alidade to point station C, record bearing BC. Carry out any other station observations required.
(4) Transit telescope, rotate alidade until circle reads bearing BC ± 180°. Check that station C is bisected by central vertical cross-hair of telescope reticule.
(5) Move to station C, repeat procedures.

Direct method without transiting
The same assumptions are made as before.
At station A:
(1) Point telescope on B. Record bearing AB.
(2) Make any other station observations.
(3) Point B again, check circle reading, leave alidade clamped, move to B.

At station B:
(1) Set up and level.
(2) Point A, using lower clamp and tangent screw, so that circle reads bearing AB.
(3) Turn alidade and point station C. Record circle reading, which will be equal to bearing BC ± 180°.
(4) Continue in this manner.

This method avoids transiting, but the instrument orientation is in error by 180° at every alternate station. This must be allowed for before commencing traverse computation.

Back bearing method
The same assumptions are made as before.

At station A:
(1) Point telescope on B. Record bearing AB.
(2) Make any other station observations.
(3) Point B again, check circle reading, move to B.

At station B:
(1) Set up and level.
(2) Set circle to read bearing BA. (Back bearing of AB).
(3) Point A with circle reading clamped at bearing BA.
(4) Turn alidade and point station C. Record bearing BC.
(5) Repeat procedures.

6·1·6   *Error in angular measurement*

The angular misclosure is determined by commencing on a fixed bearing and closing onto a fixed bearing or, in a closed polygon, by applying the angle summation test.

Provided gross error is not suspected then the angular misclosure may be distributed uniformly around all the stations of the traverse.

The permissible angular misclosure is stated in the form $A\sqrt{n}$, where n is the number of stations in the traverse. The value of A may vary from $3''$ to $1\cdot5'$, according to the precision of the equipment and the methods used.

In a long traverse, azimuth check should be obtained by astronomical observations at about every eight stations or so. An alternative check is to use a gyroscope theodolite for check azimuths.

6·2   Linear measurement

Traverse leg lengths may be measured by direct methods (chain, band, tape), by optical methods, or electromagnetic methods, the choice depending upon the particular circumstances. Whichever method is used the appropriate corrections to measured lengths must be applied.

As a check on gross error, each leg should be measured twice, but the second check measure may be of a lower standard of accuracy than the initial measure used in computing the traverse.

The linear misclosure may be indicated by stating the ratio of the total linear misclosure to the total length of the traverse, expressed as a fraction. The permissible linear misclosure should preferably be stated in the form $L\sqrt{M}$, where M is the total length of the traverse in metres. Appropriate values for L vary from 0·003 for precise work to about 0·02 for very low order work.

6·3   Traverse computation and co-ordinates

Stages in computing a traverse:
- (a)   establish bearings of the first and last lines of the traverse
- (b)   compute the bearings of the intermediate lines
- (c)   adjust the bearings of the intermediate lines
- (d)   compute the partial easting and northing of each line
- (e)   check computation of partial eastings and northings
- (f)   adjust partial eastings and northings
- (g)   determine adjusted co-ordinates.

In a closed-loop included angle traverse the *angles* may be adjusted (after checking by the angle summation test) and the adjusted bearings determined without further correction.

6·3·1   Co-ordinates

In a plane rectangular co-ordinate system, the position of a point is specified by stating its perpendicular distances from the two axes of the system. The N/S axis (*reference meridian*) and the E/W axis (*reference latitude*) intersect at the *origin* of the system.

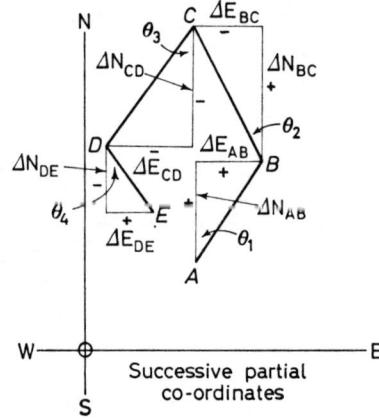

Successive partial co-ordinates

The *easting* of a point is its distance east (or west) from the reference meridian, its *northing* is its distance north (or south) from the reference latitude.

The easting and northing of a point are termed its *co-ordinates.* Easting is always stated *before* northing. In symbols, the co-ordinates of a point A may be stated as (E_A, N_A). Distances east or north are reckoned as *positive*, distances west or south are reckoned as being *negative*.

Where a line joins two stations A and B, the *partial easting* of line AB is the difference between the easting of A and the easting of B. Symbol ΔE_{AB}. Similarly, the *partial northing* of line AB is equal to the difference between the northing of A and the northing of B. Symbol ΔN_{AB}.
Partial easting and northing of a line are collectively termed its *partial co-ordinates.*

6·3·2   Calculation of partial co-ordinates

For a line of length l, and quadrant bearing θ (quadrant bearing from the reference meridian):
$$\text{partial easting} \quad = \quad l \times \sin\theta,$$
$$\text{partial northing} \quad = \quad l \times \cos\theta.$$

6·3·3   Calculation of co-ordinates

Where a line joins two points A and B, the co-ordinates of A and the partial co-ordinates of AB being known:

Easting of B $\quad=\quad$ easting of A + partial easting of AB,

and $\quad$ Northing of B $\quad=\quad$ northing of A + partial northing of AB,

or $\qquad\qquad\qquad\qquad E_B = E_A + \Delta E_{AB}$

and $\qquad\qquad\qquad\qquad N_B = N_A + \Delta N_{AB}$

If the co-ordinates of the first station of a traverse are known, the co-ordinates of subsequent stations are determined by successive addition of the partial co-ordinates of the lines of the traverse.

6·3·4   Line bearing and length from co-ordinates

Given the co-ordinates of stations A and B, the quadrant bearing (θ) and length (l) of the line AB being required:
$$\tan\theta = (E_B - E_A)/(N_B - N_A) = \Delta E_{AB}/\Delta N_{AB}$$
$$\therefore \quad \theta = \tan^{-1}(\Delta E_{AB}/\Delta N_{AB}).$$

$$\text{Length AB} = \sqrt{\{(\Delta E_{AB})^2 + (\Delta N_{AB})^2\}}$$

$$\text{or} \ \ l = \Delta N_{AB}\sec\theta = \Delta E_{AB}\operatorname{cosec}\theta.$$

6·3·5   Co-ordinates of the point of intersection of two lines

Given lines AC and BC, respective bearings α and β. The co-ordinates of A and B known, the co-ordinates of C required:
$$\Delta E_{AC} = (\Delta N_{AB} - \Delta E_{AB}\cot\beta)/(\cot\alpha - \cot\beta).$$

$$\Delta N_{AC} = (\Delta E_{AB} - \Delta N_{AB}\tan\beta)/(\tan\alpha - \tan\beta).$$

For ΔE_{BC} and ΔN_{BC}, change β in each top line to α.

6·3·6   Transformation of co-ordinates

Given points A and B, co-ordinates in the original system being (E_A, N_A) and (E_B, N_B), length of AB being l and its bearing θ.

The co-ordinate axes to be rotated through an angle α, the new bearing of AB being $\beta = \theta - \alpha$. The new co-ordinates of A given as (E'_A, N'_A).

Required to find the new co-ordinates of point B:
$$E'_B = E'_A + l\sin\beta = E'_A + (\Delta E_{AB}\cos\alpha - \Delta N_{AB}\sin\alpha)$$

and $\qquad\qquad N'_B = N'_A + (\Delta N_{AB}\cos\alpha + \Delta E_{AB}\sin\alpha)$

If scale factor k required for change of units, interpose k in front of the bracketed expressions.

6·3·7   Adjustment of bearings/angles

If the angular misclosure in bearing is e'', there being n bearings to be adjusted, the first bearing is adjusted by e''/n, the second by $2e''/n$, and so on, the last being adjusted by $ne''/n = e''$.

If the traverse is a closed loop, included angles measured, angular misclosure e'', and n stations, each angle is adjusted by e''/n.

6·3·8   Check computation of partial co-ordinates

The only valid check on the computation of partial co-ordinates is a complete re-computation using a different method from the original calculation.

Method 1: Rough check — A computation by machine or logarithms may be roughly checked by the use of traverse tables.

Method 2 — A machine computation may be checked by a re-computation using logarithms, or vice-versa.

Method 3: Auxiliary bearings — If 45° is added to each bearing, and each line length is divided by $\sqrt{2}$, 'modified' partial co-ordinates may be computed for each line.

Modified partial easting, $S = l \sin(\theta + 45°)/\sqrt{2}$,
Modified partial northing, $C = l \cos(\theta + 45°)/\sqrt{2}$,

then
$$S - C = l \sin\theta = \Delta E$$
$$S + C = l \cos\theta = \Delta N$$

The computation of $(S - C)$ and $(S + C)$ for each line provides a complete check on the original computation of ΔE and ΔN.

Alternatively, $\Delta E + \Delta N = 2S$, and $\Delta N - \Delta E = 2C$, and either of these may be used instead of calculating both S and C.

For a rapid check on the whole traverse,
$$(\Sigma\Delta N + \Sigma\Delta E)/\sqrt{2} = \Sigma l \sin(\theta + 45°),$$
and
$$(\Sigma\Delta N - \Sigma\Delta E)/\sqrt{2} = \Sigma l \cos(\theta + 45°)$$

The auxiliary bearing method is to be preferred since all bearings are completely changed — with no change of bearings in the other methods it is possible for the same mistakes to recur.

6·3·9   Adjustment of partial co-ordinates

Where a traverse is run between two fixed stations, the totals of the partial co-ordinates (with due regard to sign) should equal the differences between the co-ordinates of the fixed stations.
Similarly, where a traverse is run as a closed loop which commences and closes on the same station, the partial co-ordinates should sum to zero.

Neither of these conditions are fulfilled in practice, due to inevitable small errors, and the closing errors in the partial easting and partial northing sums must be distributed round the traverse. The partial co-ordinate misclosures are used to determine the linear misclosure of the traverse as shown in the example alongside.

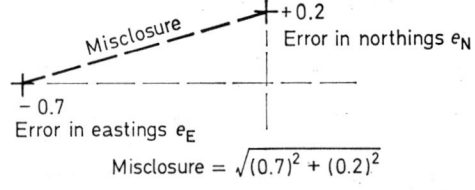

Misclosure $= \sqrt{(0.7)^2 + (0.2)^2}$

A variety of methods are available for partial co-ordinate adjustment, the two most commonly used being (a) Bowditch's Method, and (b) Transit Method.

Bowditch Method
$$\text{Correction to partial easting of a side} = \frac{\text{misclosure in partial eastings x side length}}{\text{total length of traverse sides.}}$$

$$\text{Correction to partial northing of a side} = \frac{\text{misclosure in partial northings x side length}}{\text{total length of traverse sides.}}$$

Transit Method
$$\text{Correction to partial easting of a side} = \frac{\text{misclosure in partial eastings x partial easting}}{\text{arithmetical sum of partial eastings}}$$

$$\text{Correction to partial northing of a side} = \frac{\text{misclosure in partial northings x partial northing}}{\text{arithmetical sum of partial northings}}$$

Both methods alter the line lengths and the adjusted bearings, but the Transit method disturbs the bearings less than the Bowditch method does. The Bowditch method is generally preferred by reason of its simplicity.

6·4   Checks on traverse survey

Fieldwork should be checked before leaving site. On occasion it may be impracticable to acquire all the desired data on site and office check methods may be of value.

6·4·1   Field checks

Linear measurement – All line lengths should be measured twice, with a change of method, to detect gross error.

Angular measurement – Angle measurement should be arranged in a manner which will detect gross error before leaving the site.

Traverse between fixed points – Traverse should commence and terminate at lines of known fixed bearing, the bearings being carried through and compared with the known closing bearing.

Closed-loop traverse – If direct bearing measurement used, the bearings of the first and the last lines should be measured at the first station and a check obtained as above.
If included angle measurement used, then

$$\text{sum of internal angles} = (2n - 4) \text{ x } 90°, \text{ or}$$
$$\text{sum of external angles} = (2n + 4) \text{ x } 90°,$$

where n is the number of sides or angles.
If deflection angle measurement used, the sum of the deflection angles should be $360°$.

6·4·2   Office checks

Error in length of one line – If the figure closes, determine the closing error *and its direction*. If the misclosure line is parallel to one side of the traverse, it is likely that the error lies in that side. (*see* graphic adjustment of plane table traverse.) If several lines are parallel to the misclosure line, re-measure all these lines.

Error in the angle at one station — If the figure is closed, plot the traverse and determine the closing error as before. Draw the perpendicular bisector of the misclosure line and check whether, when produced, it cuts one of the traverse stations. If one station is intersected by the bisector produced then the angular error is at that station.

Error in bearing at one station — If the traverse closes onto fixed points, plot or compute the traverse from each end in turn. If the two plots coincide at one station, then the bearing error is at that station. If there is more than one bearing in error these cannot be located.

6·4·3   Omitted measurements in closed traverse

On occasion line lengths or bearings cannot be obtained. If the traverse is closed, and not more than two quantities are missing, these can be deduced, since

$$l_1 \sin \theta_1 + l_2 \sin \theta_2 + l_3 \sin \theta_3 + \ldots l_n \sin \theta_n = 0$$

and

$$l_1 \cos \theta_1 + l_2 \cos \theta_2 + l_3 \cos \theta_3 + \ldots l_n \cos \theta_n = 0$$

where the l terms represent the traverse sides and the θ terms represent the quadrantal bearings.

These methods should not be used unless there is no alternative, since there is no means of detecting and distributing error.

Bearing of one line missing — The bearing θ required for the line joining two stations A and B.

$$\tan \theta = \Delta E_{AB}/\Delta N_{AB} = (E_B - E_A)/(N_B - N_A) = \frac{\text{sum of other partial eastings}}{\text{sum of other partial northings}}$$

Length of one line missing — The length required of the line joining two stations A and B. Bearing of AB is θ.

$$\text{Length AB} = \Delta E_{AB}/\sin \theta = \Delta N_{AB}/\cos \theta$$

Length and bearing of one line missing — Use the two methods shown above.

Lengths of two lines missing — The lengths required for lines l_1 and l_2, their respective bearings being θ_1 and θ_2. If the sum of the partial eastings of all the remaining lines is P, and the sum of all the partial northings of the remaining lines is Q, then

$$l_1 \sin \theta_1 + l_2 \sin \theta_2 = P,$$
$$l_1 \cos \theta_1 + l_2 \cos \theta_2 = Q,$$

and these two simultaneous equations may be solved to find the two unknowns.

Length of one line and bearing of another line missing. — Use the same equations as shown immediately above.

6·5   Sources of error in theodolite traversing

Linear measurement — For sources of error in direct linear measurement refer to section 2.2 For sources of error in optical distance measurement refer to chapter 8.

Angular measurement — Errors in angular measurement may lie in
- (a)   instrumental errors
- (b)   errors and mistakes in theodolite operation and setting up
- (c)   observational errors
- (d)   natural causes.

6·5·1   Instrumental errors

Residual errors of adjustment and the non-adjustable errors can be minimised by taking all observations on both faces, and if high accuracy is demanded several zeros may be used.
For most traverse work it will be adequate to measure included angles by simple reversal, as detailed in section 3.10.1.
Direct bearing methods can be carried out on both faces.

6·5·2   Errors in theodolite operation

Defective instrument centring has serious effects on short leg traverses, but the effects are not likely to be significant in ordinary work. For very short leg traverses the 'three tripod traverse equipment' should be used, with interchangeable targets/instrument placed on pre-positioned tripods.

Defective levelling results in the vertical axis of rotation not being truly vertical. The effects on horizontal angles are not significant unless the forward and back stations vary greatly in elevation.

Slip due to inadequately tightened instrument head or tripod, and the use of the wrong *clamp or tangent screw* will both result in unreliable observations.

6·5·3   Observational errors

Inaccurate signal bisection is often caused by parallax, or by sighting too high on the target.

Non-vertical signal is particularly significant in short legs. Poles should be bisected as low down as possible. Three tripod methods of value in short leg traverses.

Reading and booking errors should be guarded against by suitable routines.

Errors due to signal displacement; if temporary signals are used they should be safeguarded and referenced to permanent marks.

6·5·4   Natural causes

Wind causes tripod vibration and disturbs plumb-bob and line.

High temperatures cause irregular refraction — keep sight rays at least 1 metre above the ground.

Haze makes signal bisection difficult and unreliable.

6·6   Compass traversing

Compass traverses may be carried out by hand-held magnetic compass or tripod-mounted compass. The principle applications are in rapid reconnaissance and exploratory surveys. General advantages lie in the simplicity and portability of the equipment together with the rapidity of the fieldwork. Compass bearings can only be read to about 10′, hence the precision is much lower than that which is attainable in theodolite traversing.

The two basic methods used are:
- (a)   Free needle, also known as loose needle.
- (b)   Fixed needle.

6·6·1   Free needle traversing

The instrument is used as a normal magnetic compass with freely suspended needle. The instrument is placed at each station in turn and the magnetic bearing of each line obtained *from both ends of the line.*

If the forward and back bearing of a line do not differ by 180° it indicates that there is local attraction present at one or both ends of the line. (It is possible to occupy every second station only, but this is not recommended in view of local attraction effects.). When bearings have been corrected for local attraction they are used to plot or compute the traverse in the same way as in theodolite traversing.

Advantages of the free needle method:
- (a)   speed of fieldwork
- (b)   each line is independent and errors tend to compensate
- (c)   bearing of a line can be measured at any point in the line length.

6·6·2   Fixed needle traversing

The instrument is placed at each station in turn, and the back bearing and forward bearing from the station observed. The difference between these bearings gives the included angle at the station, and this is free from local attraction effects.

The included angles obtained are used in the same way as in theodolite work. The method is more accurate than free needle surveying, unless the traverse has a large number of very short lines.

6·6·3   Linear measurement in compass traversing

In compass traversing speed is more important than accuracy. Chain measurement is adequate, and in some applications pacing of distance will be sufficient.

6·7   U.K. National Grid System

U.K. mapped by Ordnance Survey, using Modified Transverse Mercator Projection. Unit of measurement is the international *metre.*

True origin is at longitude 2° W, latitude 49° N. False origin, to give all positive values for co-ordinates, is located 400 km west and 100 km north of true origin.

Position specified by letter H,N,S,J,O, or T, for 500 km square, followed by letter of alphabet (omitting I) for 100 km square, and finally easting and northing within the 100 km square designated.

7 Triangulation

A triangulation survey consists of a network of triangles in which one side length and all the angles are measured, the lengths of all the other sides being computed without further measurement. The single measured line is the *base line* of the network, it defines the scale of the network and it must be measured with very high accuracy. A *check base line* can be measured at the end of the net and its measured and computed lengths compared as a check on the work.

Triangulation is used for control frameworks of all levels of accuracy, but it is being replaced by electromagnetic trilateration in geodetic work.

Triangulation may be classified as *primary* (1st order), *secondary* (2nd order) or *tertiary* (3rd order), when concerned with national survey. In engineering applications may be described as *major* or *minor*.

Class	Side lengths	Triangle misclosure	Linear error	Remarks
1st order	30 – 100 km	Average ≯ 1″, max. ≯ 3″	1 : 50 000 –	Primary frame
2nd order	15–30 km	Average ≯ 3″, max. ≯ 5″	1 : 20 000–50 000	Breakdown
3rd order	1–15 km	Average ≯ 15″.	1 : 5 000–20 000	Breakdown and detail control.

A minor engineering triangulation might be similar to 3rd order.

7·1 Triangulation figures

Figures arranged in chains or belts, or as a complete net over an area.

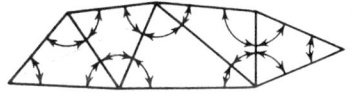

Simple chain of triangles suitable for low order work, triangles as near equilateral as possible.

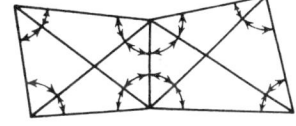

Braced quadrilaterals of four corner stations and observed diagonals are best form for high accuracy, since there are more conditions to be met in the figure. Quadrilaterals should not be long and narrow.

Centre point figures lie between the two extremes and are frequently necessary.

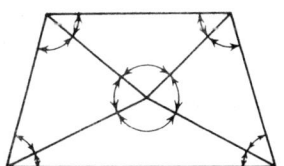

As a general rule, all triangle angles should lie between 30° and 150°.

7·2 Triangulation procedure

Order of work:
- (a) reconnaissance of area, station marking, signal erection
- (b) baseline measurement
- (c) determination of baseline azimuth or bearing and the co-ordinates of one end
- (d) measurement of all angles
- (e) computation of the triangulation.

Important points in carrying out reconnaissance:
- (a) suitable shape for triangles
- (b) station intervisibility
- (c) elimination of grazing sight rays
- (d) easy access to stations
- (e) selected points suited to later break-down for detail
- (f) baseline area as flat as possible, ends intervisible, baseline as long as possible and area suited to base extension

7·3   Base line measurement

Traditional geodetic base line measurement was by *Invar* tape in catenary, with all possible refinements, to achieve accuracy at 1: 500 000 or better. Now being replaced by electro-magnetic distance measurement methods.

Lower order base lines measured by steel tape (flat or catenary) or by traversing with subtense bar or distance wedge telemeter, or short range electronic instruments such as the Wild DI 10 T or DI 50 Distomat.

Corrections to base line measurement include:

(a)	standardisation	(e)	index error of spring balance
(b)	temperature	(f)	sag
(c)	slope	(g)	height above sea level.
(d)	change of pull (tension)		

The choice of corrections depends on methods and accuracy required.
If the base line length is reduced to mean sea level the other lines are automatically reduced to mean sea level.

Base line length may be much less than typical triangle side lengths, then the base must be extended by triangulation until one of the main triangle sides can be computed. In the example, the measured base is *ab*, extended to *c* and *d* and then to *e* and *f* and finally to the triangulation stations A and B.

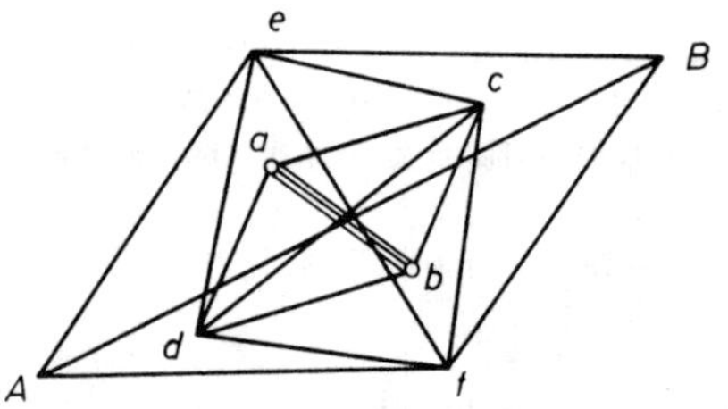

In minor triangulation it is sometimes possible to make one of the triangle sides act as the base line for the network, and this is to be preferred to using any shorter base line.

7·4   Horizontal angle measurement

Horizontal angles measured by the *direction method*. (Repetition method sometimes used with vernier instruments is more suited to small angles such as in subtense work.)

The number of rounds and zeros to be used, and the choice of instrument, depend on the accuracy demanded. Typical specifications:

Primary — 16 rounds or sets, 32 measures of each angle.
Secondary — 8 rounds, 16 measures.
Tertiary — 4 rounds, 8 measures.
A change of face and swing are made on each zero.

Errors in angular measurement have a direct effect on the fixing of the relative positions of points in the net while base measurement affects the *scale* only.

Example shows 5 directions (4 angles) to be measured from station K. Clearest target is chosen as reference object, A in example, and targets sighted in order from A, face left.

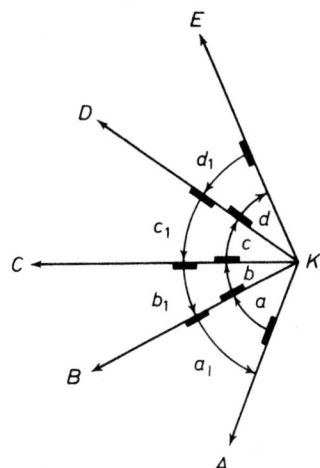

After last ray (E), transit, re-observe all targets anti-clockwise commencing at E. The round is complete when A is reached again.

If *n* sets required, change initial setting of RO by $180°/n$ for each.

If there are more than 5 or 6 directions, it is better to divide the observations into sectors of 4 or 5 rays. Each sector should be defined by a very clearly visible target, and those should be spread evenly around the horizon. The sector angles should be measured very carefully by reiteration, closing the horizon, then when sector angles accepted the angles within each sector should be measured individually or by the directions method shown above.

7·5   Triangulation computation

Order of computation after completion of fieldwork:
- (a)   computation of base line length, two separate computations for check
- (b)   abstract of angles from fieldbooks, mean values for figure angles
- (c)   adjustment of figures
- (d)   solution of triangles
- (e)   orientation of the network
- (f)   computation of co-ordinates, line bearings deduced from adjusted angles
- (g)   computation of vertical heights.

7·5·1   Figure adjustment methods

The method of adjusting triangulation figures depends upon the nature and accuracy of the triangulation concerned.
Methods available:
- (a)   least squares method
- (b)   method of equal shifts
- (c)   method of meaning results

The first method is used in primary and secondary work, where figures are often complex and high precision is essential. The second and third methods are much used in minor triangulation. Low-order work, with simple figures, it is often adequate to merely adjust the angles of each triangle to total $180°$. (add arithmetic means of the three angles, apply one third of the closing error to each angle before solving the triangle).

7·5·2   Triangle solution

Triangles solved by sine rule — $a/\sin A = b/\sin B = c/\sin C$. If the known side is a, then:
$b = a \sin B \operatorname{cosec} A; c = a \sin C \operatorname{cosec} A$.

If a simple figure adjustment to 180° has been made, several values may be computed for any one line. The several values should be meaned before calculating co-ordinates.

In geodetic work the triangles will be spherical, and angles total 180° + *spherical excess*. A spherical triangle may be computed as a plane triangle if each angle is reduced by one third of the spherical excess and this is automatically done if arithmetic means adjusted to 180°.

Spherical excess, in seconds, = area of triangle/$(R^2 \times \sin 1'')$, where R is the radius of the earth. (approximately $1''$ per 200 km^2.)

7.5.3   Network orientation

Large networks must be oriented by astronomical observations. Minor triangulations may be oriented by
  (a)   astronomical observations (sun or stars)
  (b)   compass bearing
  (c)   gyro-theodolite, or
  (d)   arbitrary direction.
The azimuth or bearing of the base line fixes the orientation of the net, but check azimuths may be required in an extensive network.

7.5.4   Co-ordinate computation

If co-ordinates of one point, such as an end of base line, are known, all other co-ordinates can be computed in the same way as in traversing. If known point co-ordinates not available, arbitrary values may be assumed for a survey of limited extent.

Plane rectangular co-ordinates used in engineering work, but geographical co-ordinates may be used initially in geodetic work.

7.5.5   Vertical height computation

Spirit levelling is preferred for station heights, but trigonometrical methods may be used in some circumstances.

7.6   Triangulation problems

The commoner problems in triangulation survey are outlined in the following sections.

7.6.1   Intersection

The method used to fix the position of a point which is inaccessible. In the figure, points 1, 2 and 3 are fixed (known co-ords), and 4 is the inaccessible point.

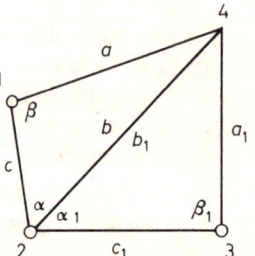

Angles α, β, α_1, β_1, measured. The lengths and bearings of c and c_1 computed from co-ordinates, then used in turn to compute lengths and bearings of a, b, b_1, a_1, taking means as necessary. Finally, co-ordinates of 4 computed
  (i)   from 1, using a and its bearing,
  (ii)   ″   2, ″ b ″   ″   ″   Mean of these values accepted.
  (iii)   ″   3, ″ a_1 ″   ″   ″

7·6·2 Resection (three point problem)

The resection method is used to fix the position of a point from angular observations to three fixed points, these points not actually being occupied.

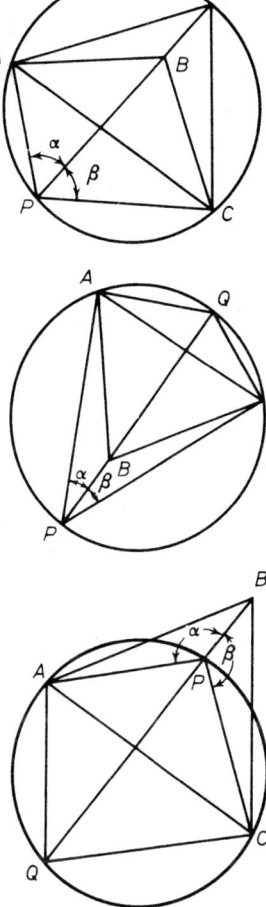

The illustration shows fixed points A, B and C, and the point to be fixed, P. Three possible layouts shown, in each the circle through A, P and C is drawn and the line PB produced to meet the circle at Q. The angles APB (α) and BPC (β) are assumed to have been observed. (Note that no solution is possible if P lies on or near the circle passing through A, B and C).

The following computation solution is the Collins Point method—the problem may be solved by other computation methods, or graphical or mechanical methods.

$A\hat{Q}C = Q\hat{P}C = \beta$, and $Q\hat{C}A = Q\hat{P}A = \alpha$.
Since A, B and C co-ordinated, deduce the bearings of AB, BC and CA.

In triangle AQC, with one side and two angles known, compute lengths AQ and QC. Compute the co-ordinates of point Q, working from A, and re-compute from C as a check.

Obtain the bearing of BQ from the co-ordinates of B and Q. Deduce the angles A$\hat{Q}$P and C$\hat{Q}$P from the bearings of their containing rays.

Solve triangle AQP to obtain length AP, and similarly triangle CQP to obtain length CP.

Compute the co-ordinates of P, working from A, and re-compute from C as a check.

(In the third figure, the method is similar but angles Q$\hat{P}$A and Q$\hat{P}$C are equal to $(180° - \alpha)$ and $(180° - \beta)$ respectively.)

It is always advisable to observe a *fourth* point and compute the bearing of the point from the calculated co-ordinates of P, then compare the observed and calculated bearings as a final check.

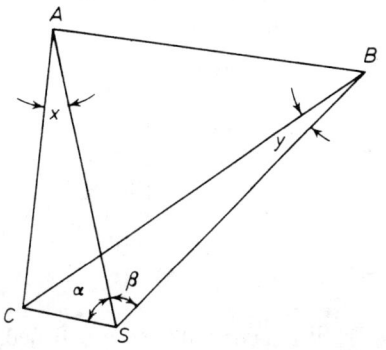

7·6·3 Satellite or eccentric station

Points A, B and C in the figure represent three triangulation stations. A and B can be occupied and angles A and B measured, but the instrument cannot be set over point C. The instrument is therefore placed over a satellite station S, as close as possible to C, and if the distance SC and the angles α and β measured, the angle A$\hat{C}$B may be deduced.

In triangle ABC, with known values for angles $C\hat{A}B$ and $A\hat{B}C$ and the known length AB, solve to obtain lengths CA and CB. (These will be approximate values, since C is unknown and A and B not adjusted.)

Using approximate values, $\sin x$ = CS sin α/AC, and if x small and stated
in seconds, x'' = CS sin α/AC. sin $1''$,
and y'' = CS sin ($\alpha + \beta$)/CB. sin $1''$.
But $B\hat{C}S$ = $180° - (\alpha + \beta) - y$,
and $A\hat{C}S$ = $180° - x - \alpha$.
∴ $A\hat{C}S - B\hat{C}S$ = $A\hat{C}B = C = \beta + y - x$

Note: High accuracy is essential, seven-figure logarithms should be used.

7·7 Figure adjustment methods

In high-accuracy work, figures are adjusted by the method of least squares. The following methods are approximate only.

7·7·1 *Method of equal shifts*

Suited to networks of simple figures including triangles, braced quadrilaterals, and simple centre-point figures. First, the figure angles are adjusted to meet the requirements of the angle conditions of the figure, then further adjustment to meet the side conditions of the figure.

Each independent figure is adjusted separately. The number of angle conditions to be satisfied by a free network is equal to $(l - l_1) - (S - S_u) + C_p + 1$, and the number of side conditions is equal to $l - 2S + 3$,

where l = number of lines
l_1 = number of lines observed one-way
S = number of stations
S_u = number of unoccupied stations
C_p = number of centre-points

Adjustment of single triangle
Only condition is that angles should sum to $180°$. Divide the difference between the sum of the arithmetic means and $180°$ by three, and this is correction to be applied to each angle.

Adjustment of braced quadrilateral
In the figure, assume all numbered angles measured, and these symbols represent the arithmetic means of the observed angles.

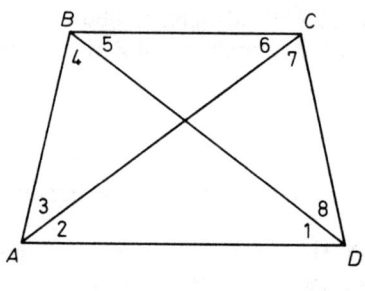

Angle conditions:
$1 + 2 + 3 + 4 - 180° = 0$ (i)
$3 + 4 + 5 + 6 - 180° = 0$ (ii)
$5 + 6 + 7 + 8 - 180° = 0$ (iii)
$1 + 2 + 7 + 8 - 180° = 0$ (iv)
$1 + 2 \ + \ 3 + 4 \ + 5 + 6 + 7 + 8 - 360° = 0$ (v)
$(1 + 2) - (5 + 6) = 0$ (vi)
$(3 + 4) - (7 + 8) = 0$ (vii)

If (v), (vi) and (vii) are satisfied, the others must be satisfied, therefore the other conditions are *redundant,* the figure has only three angle conditions to meet, numbers (v), (vi) and (vii).

First angle correction — Satisfy (v) by summing angles 1 to 8 and correct each by one eighth of the difference between the sum and 360°. Let the corrected angles be $1', 2', 3', \ldots 8'$.

Second angle correction — Apply equal corrections to $1', 2', 5'$ and $6'$, to satisfy equation (vi). Similarly, apply equal corrections to $3', 4', 7'$ and $8'$, to satisfy equation (vii).
Let the corrected angles be $1'', 2'', 3'', \ldots 8''$.
All angle conditions satisfied, since $1'' + 2'' + 3'' + 4'' + 5'' + 6'' + 7'' + 8'' - 360° = 0$.

Side condition — The satisfaction of the angle conditions does not guarantee the figure is correct geometrically — the sides must be involved also.
If length AB known, then CD can be obtained in two ways.

$$BC = AB. \sin 3/\sin 6, \quad CD = BC. \sin 5/\sin 8 = AB. \sin 3. \sin 5/\sin 6. \sin 8.$$
and $\quad AD = AB. \sin 4/\sin 1, \quad CD = AD. \sin 2/\sin 7 = AB. \sin 4. \sin 2/\sin 1. \sin 7.$

If both values of CD equal, then

$$\sin 1. \sin 3. \sin 5. \sin 7 = \sin 2. \sin 4. \sin 6. \sin 8, \text{ and using logs,}$$
$$\Sigma \log \sin \text{ odd angles} = \Sigma \log \sin \text{ even angles, or using } left \text{ and } right,$$
$$\Sigma \log \sin l = \Sigma \log \sin r.$$

The figure angles must be further adjusted to meet this condition.

Third angle correction — Find difference between left and right sides. Take out the difference for 1 second for each log sine, sum these differences for 1 second, Σd. If 1 second correction were applied to each angle, the sum of the log sines would be altered by Σd.
Divide the total correction required by Σd, the result is the number of seconds to be applied to each angle.
Corrections must be applied positively to the angles whose log sine sum is the lesser, and negatively to the others.

Note: when angle lies between 90° and 180° its log sine is positive, but the difference for 1 second is *negative*. Work in decimals of a second but round angle off to the nearest second. As final check, look out log sines of final adjusted angles and check for agreement.

Adjustment of centre point figure
The figure shows pentagon with centre point, all numbered and lettered angles observed.

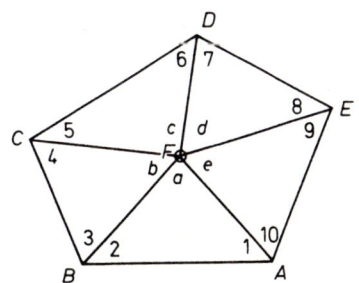

Angle conditions
$$1 + 2 + a - 180° = 0 \quad \text{(i)}$$
$$3 + 4 + b - 180° = 0 \quad \text{(ii)}$$
$$5 + 6 + c - 180° = 0 \quad \text{(iii)}$$
$$7 + 8 + d - 180° = 0 \quad \text{(iv)}$$
$$9 + 10 + e - 180° = 0 \quad \text{(v)}$$
and, $\quad a + b + c + d + e - 360° = 0 \quad \text{(vi)}$

First angle correction — Satisfy equations (i) to (v) by distributing the misclosure in each triangle equally around the angles of the triangle. Let centre corrected angles be a', b', c', d' and e'.
Let outer adjusted angles be $1', 2', \ldots \ldots 10'$.

Second angle correction — Correct centre angle equally to satisfy eqn. (vi), and then $a'' + b'' + c'' + d'' + e'' - 360° = 0$.
Adjust numbered angles in each triangle by a correction equal to, but of opposite sign from, that applied at central angle of the triangle. This will maintain equations (i) to (v).

Side condition – One condition only, as before, Σ log sin odd angles equal to Σ log sin even angles. Apply third angle correction in the same way as before.

7.7.2 Method of meaning results

Commencing from the known base, adjust each triangle to exactly 180°. Compute the triangle sides and the co-ordinates of each point, triangle by triangle, working through the network.

If two values for the length of a line are obtained, accept the mean of these, and similarly if two sets of co-ordinates are computed for a point take the mean of these values.

Computation may be speeded-up if a formula is used to calculate the co-ordinates of the third point in any triangle in which the co-ordinates of the other two points and the angles are known. In the figure, the co-ordinates of A and B are known, and the base angles α and β.

Then $E_C = \dfrac{E_A \cot \beta + E_B \cot \alpha - N_A + N_B}{(\cot \alpha + \cot \beta)}$

and $N_C = \dfrac{N_A \cot\beta + N_B \cot\alpha + E_A - E_B}{(\cot \alpha + \cot \beta)}$

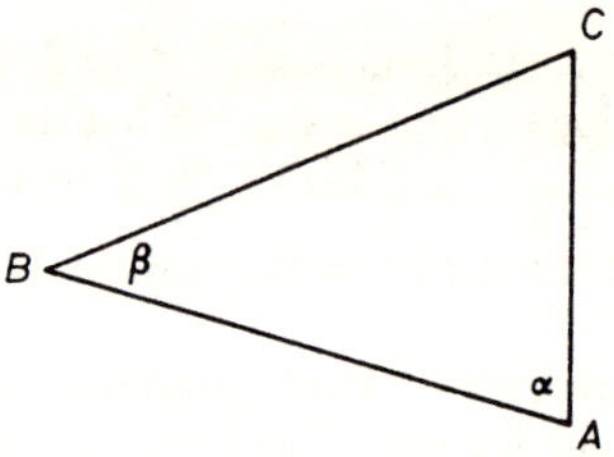

Note: It is not necessary to adjust the angles around a centre point or fixed angle – all triangles are adjusted simply to 180°.

8 Optical Distance Measurement

Principal branches:
- (a)   tacheometry, also known as tachymetry
- (b)   subtense measurement
- (c)   rangefinder—telemeter methods.

All these methods may be used to fix the position of a distant point (either a survey station or a detail point) in both height and horizontal distance by optical measurement instead of direct measurement and levelling. The horizontal direction to the point is obtained by normal horizontal bearing observations.

All methods are based on establishing either an isosceles or right-angled triangle and solving these from the observed data.

8·1   Tacheometry

In all forms of tacheometry a theodolite or an instrument of modified theodolite type, is placed over a ground station of known height and plan position and then observations are made on a suitable graduated staff placed at the target point.

As distinct from levelling, it is important that the theodolite horizontal axis height above the ground station be measured and recorded.

8·1·1   Tangential measurement

Normal theodolite is used, and a graduated staff is held *vertically* on the target point. The telescope is pointed in turn on two separate distinct readings on the staff, A and B, and the vertical angles ϕ and θ noted together with the distances s and m. For long range work, A and B could be special targets such as those used on target levelling staves.

The required measurements are:

D = horizontal distance,
Δh = difference in height between station point and target point P.

then

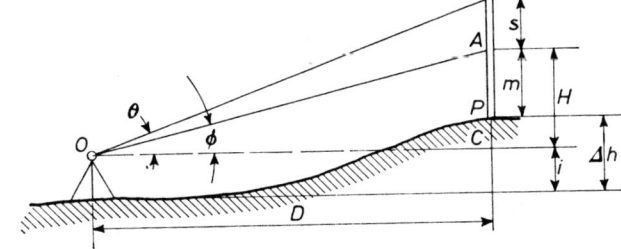

$$CA/OC = \tan \phi$$
$$\therefore \qquad CA = OC \tan \phi$$
and similarly $CB = OC \tan \theta$.

and
$$s = CB - CA$$
$$= OC \tan \theta - OC \tan \phi$$
$$= D (\tan \theta - \tan \phi)$$

therefore $D = s/(\tan \theta - \tan \phi)$, required horizontal distance.

If i is the measured instrument horizontal axis height above ground point,
then $\quad \Delta h + m = H + i$
$$\therefore \Delta h = H + i - m$$

In addition,

$$H/D = \tan \phi$$

$$\therefore \ H = D \tan \phi$$

and $\qquad\qquad \Delta h = D \tan \phi + i - m$, required height difference.

Note that marks A and B should be as far apart as is practicable.

Application: the method is not suited to general detail survey, but may be found useful occasionally.

8·1·2   Stadia measurement, or ordinary tacheometry

Normal theodolite with the usual horizontal stadia lines on the diaphragm reticule is used, and a graduated staff is held on the target point.

As shown in Section 3·1·7, if theodolite telescope is pointed on a staff which is held at right angles to the collimation line, then the distance d in the figure is equal to $(Ks + c)$, where K is the stadia multiplying constant of the instrument, s is the staff intercept, and c is the instrument stadia additive constant.

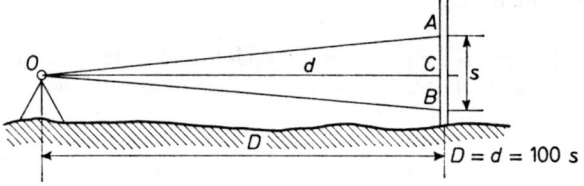

Since all modern instruments are made with K exactly equal to 100, and c so small that it may be ignored, it is assumed in the following sections that d is exactly equal to 100 x s.

Vertical staff tacheometry
This is the method most widely used in ordinary tacheometry today. The staff is held *vertically* on the target point, and the telescope directed as shown. Readings are taken at A, B and C (the stadia hairs and central cross-hair) and the vertical angle β read from the vertical circle. As before, the quantities required are D and Δh.

The slope distance is $d = 100$ x s_1, and in triangles $A_1 AC$, $B_1 BC$, the angles $AA_1 C$ and $BB_1 C$ may be taken as right angles, therefore

$$s_1 = s \cos \beta$$

and $d = 100$ x $s \cos \beta$,

then $D/d = \cos \beta$,

$\therefore D = d \cos \beta$
$\qquad = 100 \ s \cos^2 \beta$, required horizontal distance.

Note: If the theodolite vertical circle reads the zenith distance z,
then $\qquad D = 100 \ s \sin^2 z$.

If i is the measured instrument horizontal axis height above ground point,

$$\Delta h + m = H + i,$$
$$\therefore \quad \Delta h = H + i - m$$

and
$$H/d = \sin \beta$$
$$\therefore \ H = d \sin \beta$$
$$= 100 \ s \sin \beta \cos \beta$$

and $\qquad \Delta h = 100 \ s \sin \beta \cos \beta + i - m$, required height difference.

Note: If the theodolite vertical circle reads the zenith distance z,
then $H = 100 \ s \sin z \cos z$.
It should be noted also that $\sin \beta \cos \beta = \frac{1}{2} \sin 2\beta$, and this form is sometimes preferred.

Reduction methods: Special tacheometric tables or slide rules, or diagrams.

Plotting methods: Polar co-ordinates (protractor and scale) or, for traverse stations and similar points, rectangular plane co-ordinates may be computed.

Sources of error:
(i)   Differential refraction – makes s unreliable,
(ii) Non-vertical staff – again makes s inaccurate,
(iii) Equipment faults – not generally significant today.

Accuracy: Assuming staff held properly vertical (best if staff bubble used), sight distances under about 250 metres, and vertical angles not exceeding about ±20°, then the error in any single distance measurement should be within about 1/1000 and the error in any individual height should not exceed ± 0·02m.

Special instruments: The common form today is the self-reducing diagram tacheometer, with versions made by several manufacturers. Usually these include a glass plate bearing engraved curved lines which replace the stadia lines. The distance between the curves varies with the tilt of the telescope, thus there is automatic compensation for slope.

The figure illustrates the view in a **Wild RDS** tacheometer, using an ordinary staff. The horizontal distance D is equal to 100 times the staff intercept between the outer lines. The distance between the bottom and the intermediate lines is read in the normal way as 21·7 m, then multiplied by the constant (+0·1) visible by the centre curve to give height difference of +2·17 m.

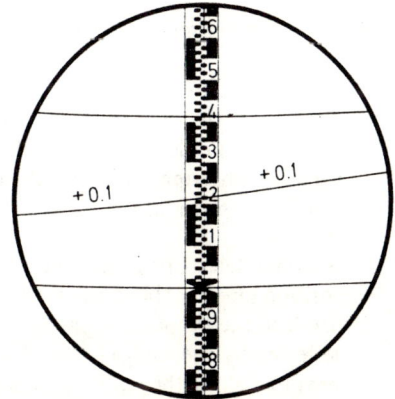

Distance = 41.3 m
Height = 0.1 x 21.7 = + 2.17 m

Another system, still used on plane table alidades, is the Beaman Stadia Arc. This has graduated arcs fixed to the vertical circle, one bearing values of $100 \sin^2 \beta$, the other $100 \sin \beta \cos \beta$. These values are read off and used with the observed staff intercept s. Not as rapid as the diagram type instruments.

Normal staff tacheometry

In this method, which is little used today, the staff is held upright on the target point and then it is leaned over (towards the theodolite) until it is at right-angles to the instrument collimation line. A sighting device is fixed to the staff for this purpose.

As before, the stadia readings are A and B, and C the middle hair point.

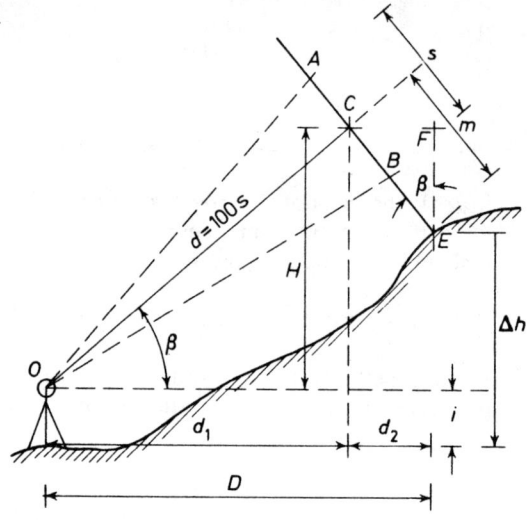

The slope distance is
$$d = 100 \times s.$$

The required horizontal distance is
$$D = d_1 + d_2$$

but $d_1 = d \cos \beta = 100 \, s \cos \beta,$

and $d_2 = CF = CE \sin \beta = m \sin \beta,$

$\therefore \quad D = 100 \, s \cos \beta + m \sin \beta.$

Note: $m \sin \beta$ is negative when β is an angle of depression.

The required height difference is Δh. From the figure,

$$H/d = \sin \beta$$
$$\therefore H = d \sin \beta$$
$$= 100 \, s \sin \beta$$

and $\Delta h = H + i - FE$
$$= H + i - m \cos \beta$$
$$= 100 \, s \sin \beta - m \cos \beta + i.$$

Reduction of observations is by ordinary trig. tables and is more cumbersome than the usual vertical staff method.

8·1·3 *Horizontal staff tacheometry, or distance wedge tacheometry*

Normal theodolite used, but with a special prism attachment placed over the objective. The attachment holds a glass wedge across the middle third of the objective and this deflects light rays entering the wedge through an angle whose tangent is 0·01, as shown in the figure.

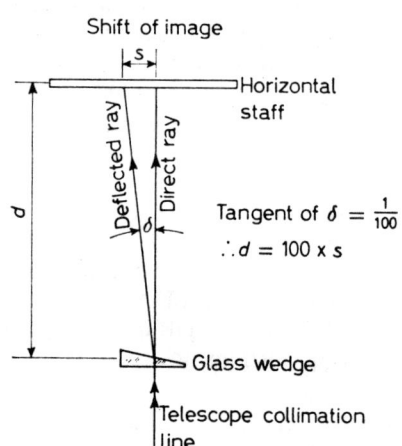

When the telescope is directed on a staff held in a horizontal position directly above the target point, two staff images are visible in the eyepiece — one as viewed direct and one as deflected by the wedge prism. The distance to the staff, measured along the collimation line, will be equal to 100 times the distance between the two points on the staff.

In practice, one staff edge is graduated and the other carries an index and vernier scale. When the prism is properly adjusted the vernier scale appears to have been shifted along the graduations and the reading at the vernier gives the amount of shift and hence the slope distance d.

For greater reading precision a plane parallel plate micrometer is built-in, thus permitting a parallel displacement of the deflected image so as to bring the vernier scale into coincidence with the main scale on the staff. The parallel plate is operated by a micrometer head graduated from 1 to 10, the figures indicating the numbers of centimetres to be added to the staff reading.

In the example reading shown, the vernier has been brought into coincidence. The vernier index is just past the eighth sub-division between 5 and 6, and this is read as 58 metres. The fourth vernier division coincides with a main scale division and this is read as 0·4 metres. The micrometer shows 4, read as 0·04 metres

The reading obtained is the slope distance d. The required values are D, horizontal distance, and Δh, height difference.

$D = d \cos \beta$, but the *preferred method* is to calculate the correction $d(1 - \cos \beta)$ and deduct this from the observed value of d. Manufacturers supply tables for this method.

$$H = d \sin \beta, \text{ and}$$

$$\Delta h = H + i - m$$
$$= d \sin \beta + i - m.$$

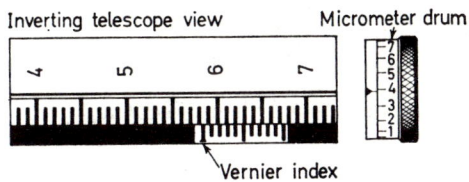

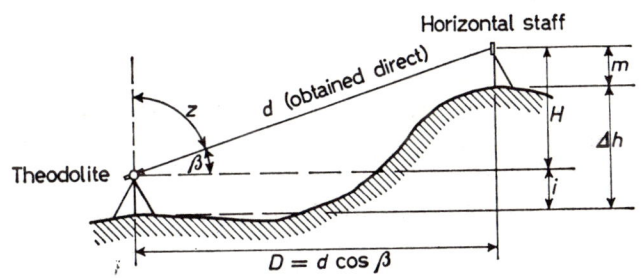

This is simplified if the staff is always set up at the same height as the theodolite horizontal axis.

Sources of error: (i) incorrect telescope focusing
(ii) staff not exactly at right angles to the collimation line
(iii) staff not levelled laterally.

Accuracy: Manufacturer's claims range from 1/5000 to 1/10 000. Best results are obtained by making several coincidence settings with the vernier and micrometer and taking the mean of the readings obtained. Wild estimate a p.s.e. of 1/5 000 if three settings are meaned with their DM1 attachment, but 1/7 500 for six.

Working range: minimum 10 m, maximum 115 − 150 m depending on the equipment.

Special instruments: A variety of self-reducing instruments are available. These incorporate double wedges which rotate as the instrument telescope tilts and thus vary the refraction angle and automatically reduce the observed distance to horizontal. All models give horizontal distance directly without calculation and some also give height difference directly. In some types the vertical circle carries a scale of tangent values and these are applied to the observed horizontal distances in order to give the height differences.

8·2   Subtense measurement

The modern version of this method uses a normal theodolite, preferably universal type (direct reading to
1″), with a 2 metre precise subtense bar set horizontally above the target point and aligned at right angles
to the direction to theodolite.

The horizontal angle subtended by the bar at the
theodolite station being measured on the horizontal
circle of the theodolite,
then

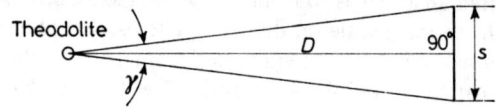

$$D = \tfrac{1}{2}\, s \cot \tfrac{1}{2}\gamma$$

Parallactic angles in subtense work being very small, it is valid to use the approximations for small angles,
then
$$D = s/2 \tan \tfrac{1}{2}\gamma$$
$$= s/\gamma'' \sin 1''.$$

With a 2 metre bar, the expressions above reduce to $D = \cot \tfrac{1}{2}\gamma$ and $D = 2/\gamma'' \sin 1''$,
the first being suited to tables supplied by manufacturers.

If heights are required, then $H/D = \tan \beta$,
$$\therefore \qquad H = D \tan \beta,$$
$$\text{and} \quad \Delta h = H + i - m$$
as in earlier methods.

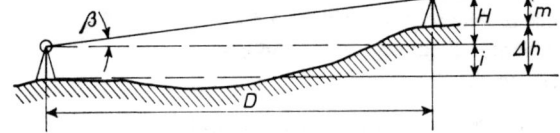

> *Precision bar*—Bar is always supported on a tripod, and fitted with aiming target at each end of the
> precise 2 m span. A sighting device is fitted so that the bar may be laid at right angles to the line to
> the instrument station.

> *Sources of error*—(a) variation in bar length, s. (b) faulty bar alignment (c) observational error in
> angle γ. It can be demonstrated that with modern equipment, and using reasonable care, the first
> two sources can be ignored in most work, and the significant source lies in the measurement of
> parallactic angle γ. It is general practice to keep the standard error in γ within ± 1″, and recent
> research would indicate that to achieve this at least eight measures of the angle are necessary with a
> 1 second theodolite.

8·2·1   Accuracy of subtense measurement

If
$$D = s/2 \tan \tfrac{1}{2}\gamma, \text{ and } \tan \tfrac{1}{2}\gamma = \tfrac{1}{2}\gamma \text{ radians,}$$
then
$$D = s/(2 \times \tfrac{1}{2}\gamma)$$
$$= s/\gamma.$$

Differentiating,
$dD/d\gamma = - s/\gamma^2$, then substituting e_D and e_γ for finite small errors in D and γ,

$$e_D = - se_\gamma/\gamma^2, \text{ and substituting } \gamma = s/D \text{ and neglecting signs,}$$
$$e_D = D^2\, e_\gamma/s$$
$$= D^2\, e_\gamma'' \sin 1''/s.$$

Assuming $e_\gamma = \pm 1''$, $s = 2$ metres,
then $e_D = \pm D^2/(2 \times 206\,265)$ metres, or the error in distance measurement will increase as the square
of the distance measured.

For practical work, the error in distance may be taken as $\pm D^2/(4 \times 10^5)$ metres.

Proportional standard error in distance —The ratio of the errors in distance and angle measurement may be deduced as $e_D/D = e_\gamma/\gamma$, where e_γ and γ are stated in the same units. It is common practice to aim for a p.s.e. of 1/10 000 in subtense work, then $e_D/D = 1/10\ 000$. If the approximate value for e_D stated earlier is substituted, then $D^2/(D \times 4 \times 10^5) = 1/10\ 000$, and $D = 40$ metres.

Thus a p.s.e. of 1/10 000 cannot be achieved over distances in excess of approximately 40 metres. (41·25 m if the more exact value for e_D is used.)

If the required p.s.e. is reduced the range can be increased, the maximum range for a p.s.e. of 1/5 000 being approx. 80 metres. If better p.s.e. values are demanded on longer lines then more complex field arrangements are needed. The common field methods are outlined in the following sections.

8·2·2 Bar at centre of line method

When the bar is placed approximately at the centre of the line, and the angles γ_1 and γ_2 measured, then

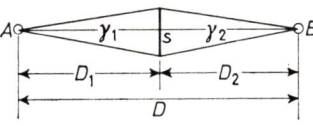

$$D = D_1 + D_2$$
$$= \tfrac{1}{2}s(\cot \tfrac{1}{2}\gamma_1 + \cot \tfrac{1}{2}\gamma_2)$$

Making the usual small angle approximations, then

$$D = s(1/\gamma_1{}'' + 1/\gamma_2{}'')/\sin 1''$$

Assuming a 2 metre bar, these may be expressed as

$$D = \cot \tfrac{1}{2}\gamma_1 + \cot \tfrac{1}{2}\gamma_2, \text{ and}$$
$$D = 2(1/\gamma_1{}'' + 1/\gamma_2{}'')/\sin 1''$$

the former being suited to the special subtense tables.

If $D_1 \approx D_2$, then $\gamma_1 \approx \gamma_2$, and $e_{\gamma 1} = e_{\gamma 2} = e_\gamma$, and

$$e_D = \pm D^2\, e_\gamma{}'' \sin 1''/2s\sqrt{2},$$
and taking e_γ as $1''$, s as 2 metres,
$$e_D \approx \pm D^2/(2\cdot8 \times 4 \times 10^5) \text{ metres.}$$

Note that this is an improvement of $1/2\cdot8 \approx 1/2\sqrt{2}$ over the direct method. For a p.s.e. of 1/10 000, line length D should not exceed approx. 117 metres.

8·2·3 Sub-divided sections method

If the line length is divided into n equal sections, then each section may be measured by the direct method and total line length is the sum of the separate sections.

With the usual assumptions,
then $D = \cot \tfrac{1}{2}\gamma_1 + \cot \tfrac{1}{2}\gamma_2 + \ldots \cot \tfrac{1}{2}\gamma_n$
$$= 2(1/\gamma_1{}'' + 1/\gamma_2{}'' + \ldots 1/\gamma_n{}'')/\sin 1''.$$

The error in distance may be deduced as

$$e_D = \pm D^2\, e_\gamma{}'' \sin 1''/sn^{3/2}$$
$$\approx \pm D^2/(4 \times 10^5 \times n^{3/2}) \text{ metres.}$$

This is an improvement of $1/n^{3/2}$ over the simple direct method.
For a p.s.e. of 1/10 000, $n \approx 0\cdot084\, D^{2/3}$.

8·2·4 Auxiliary base at end method

If an auxiliary base is set out at right angles to the line D, the base length may be measured by subtense bar then the length D obtained by measuring the angle γ subtended by the base d at point A.

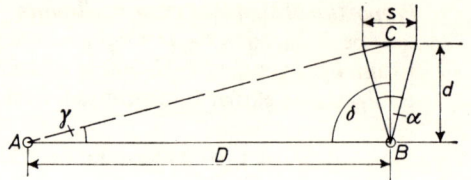

$D = d \sin (\delta + \gamma)/\sin \gamma$, where $d = \frac{1}{2}s \cot \frac{1}{2} \alpha$.
With usual small angle approximations,

$$D = s/\alpha''\gamma'' \sin^2 1''$$

and with 2 metre bar,

$$D = 2/\alpha''\gamma'' \sin^2 1''$$

For the best results, δ should be approximately $90°$, and $\alpha = \gamma$, then $d = (sD)^{1/2} = (2D)^{1/2}$.

Error in distance may be deduced as

$$e_D = \pm D\, e_\alpha'' \sin 1'' \, (2D/s)^{1/2}$$

and taking e_α as $1''$, s as 2 metres,

$$e_D = \pm D^{3/2} \sin 1''$$
$$\approx \pm D^{3/2}/(2 \times 10^5) \text{ metres.}$$

For a p.s.e. of 1/10 000, maximum length for D is 425 m, requiring an auxiliary base of 29 metres length.

8·2·5 Auxiliary base at centre method

If the auxiliary base is placed approximately at the centre of the line, then the two halves may be measured as in the last method and added to give the whole line length.

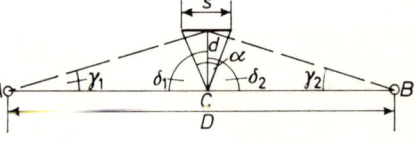

$D = AC + CB = d\,[\sin (\delta_1 + \gamma_1)/\sin \gamma_1 + \sin (\delta_2 + \gamma_2)/\sin \gamma_2]$, where $d = \frac{1}{2}s \cot \frac{1}{2}\alpha$.

If $AC \approx CB$, and $\delta_1 \approx \delta_2 \approx 90°$ and small angle approximations made,
then $D = s(1/\alpha''\gamma_1'' + 1/\alpha''\gamma_2'')/\sin^2 1''$
and with 2 metre bar,

$$D = 2\,(1/\alpha''\gamma_1'' + 1/\alpha''\gamma_2'')/\sin^2 1''.$$

For the best results $\gamma_1 \approx \gamma_2$, and also $d \approx 0.6\,(sD)^{1/2} = 0.6\,(2D)^{1/2}$

Error in distance may be deduced as
$$e_D = \pm D\, e_\alpha'' \sin 1'' \, (2D/s)^{1/2}/1.7$$
and taking e_α as $1''$, s as 2 metres,
$$e_D \approx \pm D^{3/2}/(1.7 \times 2 \times 10^5) \text{ metres.}$$

For a p.s.e. of 1/10 000, maximum length for D is 1135 m, requiring an auxiliary base length of 34 m.

8·2·6 Two auxiliary bases method

Here a short auxiliary base is set out as an extension of line D, and measured by subtense bar. A further base is set out at right angles to line D and angles γ and β measured.

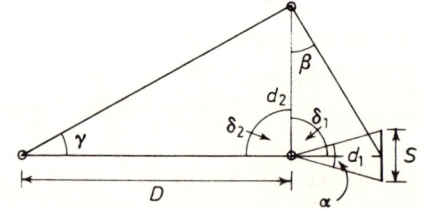

$d_1 = \frac{1}{2}s \cot \frac{1}{2}\alpha$; $d_2 = d_1 \cot \beta$; $D = d_2 \cot \gamma$;
∴ $\qquad D = \frac{1}{2}s \cot \frac{1}{2}\alpha \cot \beta \cot \gamma$.

With the usual small angle approximations, and $\delta_1 = \delta_2 = 90°$,
then $\quad D = s/\alpha'' \, \beta'' \, \gamma'' \sin^3 1''$,
and with a 2 metre bar,
$$D = 2/\alpha'' \beta'' \gamma'' \sin^3 1''.$$

For best results, $\alpha = \beta = \gamma$, then $d_1 = s^{2/3} \, D^{1/3} \approx 1.6 \, D^{1/3}$, and $d_2 = (d_1 \, D)^{1/2}$.

Error in distance may be deduced as
$$e_D = \pm D^{4/3} \, e_\alpha'' \sin 1'' \, (3)^{1/2} / s^{1/3}, \text{ and taking } e_\alpha \text{ as } 1'', s \text{ as 2 metres,}$$
$$e_D \approx \pm D^{4/3} / (1.5 \times 10^5) \text{ metres.}$$

For a p.s.e. of 1/10 000, maximum length for D is approximately 3 400 m, with $d_1 = 24$ m and $d_2 = 286$ m.

8·3 Telemetry/Rangefinders

These instruments establish a right-angled triangle which has its base at the instrument and its apex at the observed target point — a general reversal of stadia and subtense methods.

8·3·1 Rangefinders

Typical rangefinder has a short measured base AB, built into the instrument. The angle at A is fixed at $90°$ while the angle at B is varied by an adjustable prism. The angle at B is adjusted until the two images of the distant point C appear to coincide in the eyepiece at E. When coincidence is achieved the slant distance may be read directly off a graduated scale.

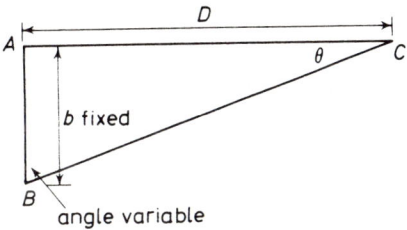

The scale gives values of $b \cot \theta$, and no angles need be measured.

The comparatively low accuracy of the rangefinder makes it suitable for reconnaissance and similar work.

8·3·2 Telemeters

These are similar to rangefinders, but the angle B is fixed and the base length is variable.

The most notable instrument of this type is the Zeiss (Jena) BRT 006 reducing telemeter. In this instrument, angle B is set at $\cot^{-1} 1/200$ so that $D = 200 \, b$, and the prism at B is adjusted by sliding it along the graduated base arm. This instrument may be set to read off slant *or* horizontal distance.

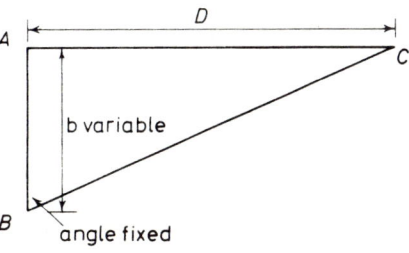

A p.s.e. of 1/1600 can be achieved at up to 60 m range without special targets. Targets are used for larger ranges, up to 180 m.

9 Areas and Volumes

9·1 Regular areas

Triangle   Area $A = \frac{1}{2}bh = \sqrt{s(s-a)(s-b)(s-c)}$
$= \frac{1}{2} ab \sin C$
$= \frac{1}{2} ac \sin B$
$= \frac{1}{2} bc \sin A$
where h = triangle height, $s = \frac{1}{2}(a + b + c)$

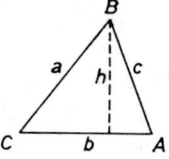

Trapezoid $A = l(a + b)/2$
where a and b are the parallel sides, l is the perpendicular distance between them.

Circle $A = \pi r^2 = \pi d^2/4$ where r = radius, d = diameter, $\pi = 3\cdot1416$.

Sector of circle $A = \pi r^2 \theta/360 = \frac{1}{2} r^2 \theta_{rad}$ where θ is the sector angle.

Segment of circle $A = r^2 (\pi\theta/360 - \frac{1}{2} \sin \theta) \simeq (3h^3 + 4c^2h)/6c$
where c is the chord length, h = segment height, θ = sector angle.

Ellipse $A = (\pi/4) ab$ where a and b are the major and minor axes.

Parabola $A = \frac{2}{3} bh$
where b affd h are the respective base widths and height of the parabola.

Curved surface of cone $A = \frac{1}{2}$ x base circumference x slant height.

Surface area of sphere $A = 4\pi r^2$.

Surface area of segment of a sphere $A = 2\pi rh$ where h = height of segment.

9·2 Irregular areas.

Where an area has irregular boundaries, the boundaries may be straightened with 'give and take' lines, then area computed by division into triangles, trapezoids, etc. Alternative methods: Grid of squares on tracing paper, computing scale, or use of planimeter.

9·2·1 Trapezoidal rule for areas.

Area BXYC in the figure.

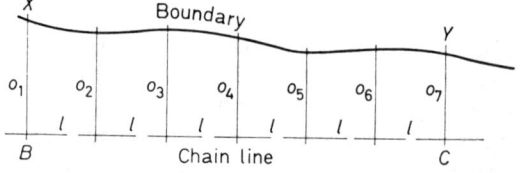

$$A = l\{\frac{0_1 + 0_n}{2} + 0_2 + 0_3 + 0_4 + \ldots + 0_{n-1}\}$$

where $0_1, 0_2, \ldots 0_n$ are offsets or ordinates, and l is the uniform distance between offsets. n may be odd or even. Resulting area is generally *less* than the true area.

9·2·2 Simpson's rule for areas.

For same area, $A = (l/3)\left[(0_1 + 0_n) + 2(0_3 + 0_5 + \ldots 0_{n-2}) + 4(0_2 + 0_4 + \ldots + 0_{n-1})\right]$. Most accurate method. n must be an *odd* number.

9·3   Cross-section areas.

Road and similar earthworks volumes obtained from the successive cross-section areas. Cross-section areas may be computed by squares, planimeter, or formulae. In the following formulae, the given data are normally h_0, formation height as measured on centre-line, b, formation width, side-slope gradients 1 in m or n, ground cross-fall gradients 1 in k or l.

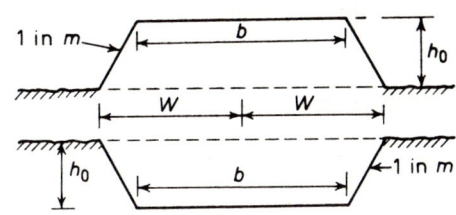

9·3·1   Level across section

Area, cut *or* fill, $A = h_0 (b + mh_0)$.

Side stake width, $W = \dfrac{b}{2} + mh$.

9·3·2   Two-level section.

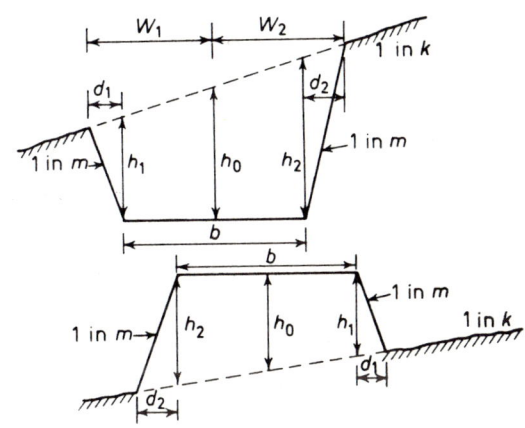

Cross-fall 1 in k.
Total area, $A = \frac{1}{2} [h_1 d_1 + h_2 d_2] + h_0 b$.

$h_1 = h_0 - b/2k; \; h_2 = h_0 + b/2k$

$d_1 = h_1 km/(k + m); \; d_2 = h_2 km/(k - m)$

9·3·3   Part cut/part fill section

Area of cut
  $= \frac{1}{2} h_2 d_2$
  $= d_2^2/2(k - m)$
  $= (b/2 + kh_0)^2/2(k - m)$

Area of fill
  $= \frac{1}{2} h_1 d_1$
  $= d_1^2/2(k - n)$
  $= (b/2 - kh_0)^2/2(k - n)$

If h_0 is positive, measured in the fill,

area of cut $= (b/2 - kh_0)^2/2(k - m)$.

Area of fill $= (b/2 + kh_0)^2/2(k - n)$

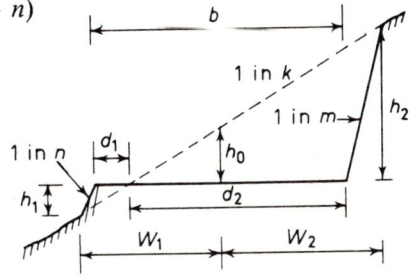

$d_1 = (b/2) - kh_0. \qquad d_2 = (b/2) + kh_0$
$h_1 = d_1/(k - n). \qquad h_2 = d_2/(k - m)$
$W_1 = (b/2) + nh_1. \qquad W_2 = (b/2) + mh_2$

9·3·4   Variable cross-fall (three level) section

Total area $= \frac{1}{2}[h_1 d_1 + h_2 d_2 + (b/2)(2h_0 + h_1 + h_2)]$
$d_1 = h_1 ml/(l - m); \; d_2 = h_2 mk/(k + m)$
$h_1 = h_0 + (b/2k); \quad h_2 = h_0 - (b/2l)$

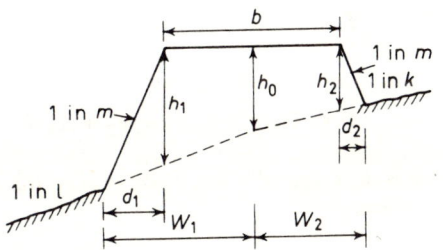

71

9·4 Areas from co-ordinates

Area of a closed traverse, *longitude method.*
Area of traverse = algebraic sum of the products of the longitude of each line and the partial northing of the line.
Longitude of a line = distance from N/S axis (reference meridian) to the centre of that line.
No sketch required. Longitudes always positive. Partial northings may be positive or negative.

9·5 Adjustment of areas measured with incorrect length chain

If A_t = true area, A_m = measured area, l = nominal chain length, $\pm\delta l$ = error in chain length,

$$A_t = A_m \left\{ 1 \pm (\delta l/l) \right\}^2$$

If error in standardisation small compared with area, % error in area is equal to approximately 2 x % error in length.

9·6 Regular volumes

Prism. Volume, $V = Ah$
Where A = cross-section area, h = perpendicular height.

Cylinder. $V = Ah = \pi r^2 h.$

Pyramid. $V = \frac{1}{3} Ah.$
Where A = base area, h = perpendicular height.

Sphere. $V = \frac{4}{3} \pi r^3.$

Prismoid. $V = (l/6)(A_1 + 4M + A_2).$
Where A_1 and A_2 are the end-face areas, M the area of the cross-section midway between the faces, l = the distance between the endfaces.

9·7 Earthwork volumes from cross-sections

Road and similar earthworks volumes obtained from successive cross-section areas.

9·7·1 Mean areas method

Successive cross-section areas $A_1, A_2, \ldots A_n$, and distance from A_1 to A_n is L.
$V = (L/n)(A_1 + A_2 + \ldots + A_n) = (L/n)\Sigma A.$ Low accuracy, volume too large.

9·7·2 End areas method

Two successive cross-section areas A_1, A_2, their distance apart l.

$$V = (l/2)(A_1 + A_2)$$

9·7·3 Trapezoidal rule

Successive cross-section areas $A_1, A_2, \ldots A_n$, at uniform distance l apart.

$$V = l\left\{ (A_1 + A_n)/2 + A_2 + A_3 + \ldots A_{n-1} \right\}$$

Result is generally *less* than the true volume. A development of the end areas method. n may be odd *or* even.

9.7.4   Simpson's rule

Successive cross-section areas $A_1, A_2, \ldots A_n$, at uniform distance l apart, and n an odd number.
$$V = (l/3)(A_1 + 4A_2 + 2A_3 + 4A_4 + \ldots 2A_{n-2} + 4A_{n-1} + A_n).$$

Note multipliers in sequence $1, 4, 2, 4, 2, 4, 2, \ldots . 2, 4, 1$.
The most accurate method of obtaining volumes.

9.7.5   Earthworks curved in plan

Pappus's theorem – The volume swept out by a constant area revolving about a fixed axis is given by the product of the area and the length of the path of the centroid of the area.

Where earthworks centre line follows circular plan, and cross-section area constant, V = cross-section area x length of arc traced out by cross-section centroid.

Section areas vary, centroid positions vary, therefore areas are corrected rather than the centroid path length deduced.
In figure, volume swept out is
$$V = A(R - e)\theta_{rad}, \text{ or } V = A(R + e)\theta_{rad}$$

if centroid beyond centre line.

Error in volume for length θR along the centre line is $\pm Ae\,\theta$, and error per unit length is $\pm Ae/R$.
Add or subtract correction to area at each section and use the usual formulae.

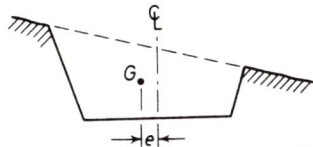

Eccentricity of centroid from ₵

Eccentricity of centroid

Two-level section $e = W_1 W_2 (W_1 + W_2)/3kA$

Part cut/part fill section,

fill area eccentricity, $e = \frac{1}{3}(W_1 + b/2 + kh_0)$

cut area eccentricity, $e = \frac{1}{3}(W_2 + b/2 - kh_0)$

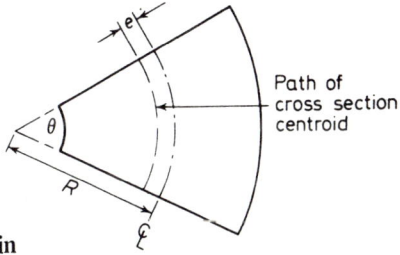

Path of cross section centroid

9.8   Adjustment of volumes measured with incorrect length chain

If V_t = true volume, V_m = measured volume, l = nominal chain length, $\pm\delta l$ = error in chain length,
$$V_t = V_m\{1 \pm (\delta l/l)\}^3$$

9.9   Volumes from spot heights

Volume of excavation obtained from grid of spot heights and treating each grid square as the top of either one square or two triangular truncated prisms.
Square prism volume = plan area x mean height of four corners.
Triangular prism volume = plan area x mean height of three corners.

Total volume obtained as:
(a) ¼ plan area of one prism x the sum of each height in turn multiplied by the number of squares in which it appears. V = (square area/4) Σhn.
or
(b) $^1/_3$ plan area of one prism x the sum of each height in turn multiplied by the number of triangles in which it appears. V = (triangle area/3) Σhn.

10 Setting Out

The operations covered by the term 'setting out' include here:

    (a)   ranging horizontal road/rail curves, circular or transition

    (b)   setting level and distance pegs for vertical curves.

10·1   Circular curves

Circular curves may be defined by either:

    (a)   the curve radius, R, in metres, or

    (b)   the degree of curvature, D, being the angle subtended at the centre of the circle by an arc of 100 metres radius.

Note that sometimes a 100 unit *chord* has been used instead of the arc. The lengths of arc and its chord differ by approximately $l^3/24R^2$, where arc length equals l.

Through chainage generally used on roads, with pegs placed at 100 m stations, these being numbered, and the distance to intermediate pegs noted as a 'plus distance', e.g. typically pegs marked 11 + 20·0, 11 + 40·0, etc., meaning station 11 plus 20·0 m, or 1120·0 m from start. Pegs placed at regular intervals on straights and curves alike, resulting in odd length arcs at beginning and end of curves.

10·1·1   Circular curve geometry

BC, beginning of curve, sometimes noted as TP1.
EC, end of curve, sometimes TP2.
PI, point of intersection of straights.
Δ, deviation angle of curve.
$180° - \Delta$, intersection angle.
T, tangent length, $= R \tan (\Delta/2)$.
ED, external distance, $= R$ exsecant $(\Delta/2)$
$$= R(\sec(\Delta/2) - 1)$$
$$= R \left\{ \frac{1}{\cos (\Delta/2)} - 1 \right\}$$
e, crown point of curve.
ef, height of long chord $= R$ versine $(\Delta/2)$
$$= R \left\{ 1 - \cos (\Delta/2) \right\}$$
df, $R \cos (\Delta/2)$, radius minus long chord
                     height.
ac, Long chord $= 2R \sin (\Delta/2)$.
L, length of curve $= R\Delta_{rad} = \pi R\Delta/180$,
    Δ stated in degrees and decimals.

Chord/arc from tangent point BC or EC -- For arc length l from a tangent point, tangent deflection angle for arc δ, central angle is 2δ, then

    2δ radians $= l/R$,

$\therefore$ δ minutes $= 206265\ l/(2R \times 60) = 1718·9\ l/R$ minutes.
Deflection angles such as δ may also be known as *tangential angles*.

10·1·2   Setting out methods

Methods available:

    (a)   deflection angles using theodolite at tangent point and measured arc/chord.

    (b)   deflection angles from both tangent points simultaneously.

    (c)   linear measurement methods.

Deflection angles and chords

Chord lengths chosen such that difference between chord/arc lengths are negligible, e.g. 3 or 4 mm. For radii in excess of 300 m, 20 m chords are suitable. Shorter radii, shorter chord.

Theodolite placed on BC, aligned along tangent, and successive chords/deflection angles set off, using the method of section 10·1·1.

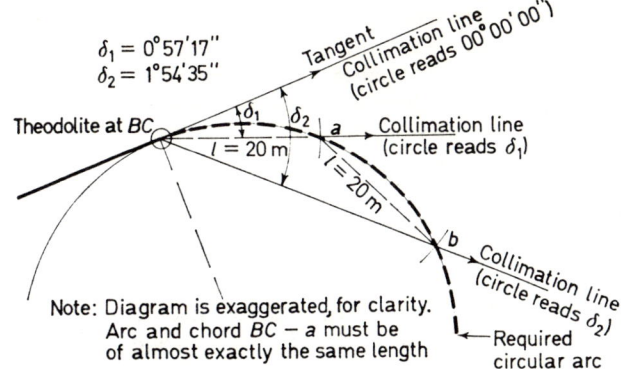

$\delta_1 = 0°57'17''$
$\delta_2 = 1°54'35''$

Note: Diagram is exaggerated, for clarity. Arc and chord $BC - a$ must be of almost exactly the same length

Successive chord/deflection angle intersections staked, at through chainage points with regular chord lengths. Sub-chords needed at BC and EC. To provide check, locate crown point and set out from both ends to the crown. Finally adjust inevitable small discrepancies.

Deflection angles with two theodolites

Compute deflection angles from both tangent points for each peg, locate by intersection of theodolite collimation lines with a theodolite on each tangent point. Useful on broken ground.

Deflection distances (linear method)

Select chord length l. Lay off along tangent from BC, swing right through arc b_1b, such that $b_1b = l^2/2R$. Lay off l on line abc_1 from b, swing through arc c_1c such that distance $c_1c = l^2/R$. Repeat as necessary, staking points b, c, d, etc.

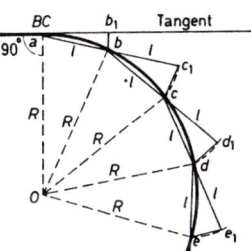

Proof: $c_1c/bc = bc/ob$, $\therefore c_1c = bc^2/ob = l^2/R$.

Offsets from long chord (linear method)

Locate BC, EC, e, and f. The perpendicular offset pq from any point p lying on the long chord at a distance x from f is equal to
$$\sqrt{(R^2 - x^2)} - R\cos(\Delta/2)$$

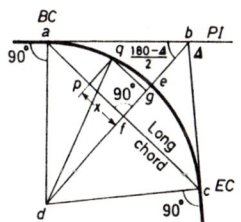

Proof: Triangle dqg, qg = x, fg = pq, then
$$df + pq = \sqrt{(R^2 - x^2)}, \therefore pq = \sqrt{(R^2 - x^2)} - R\cos(\Delta/2)$$
Alternatively, $pq = \sqrt{(R^2 - x^2)} - \{ R - \text{versine}(\Delta/2) \}$

Offsets from tangent

Locate BC and tangent alignment. The perpendicular offset pq, from any point p lying on the tangent at a distance x from BC, is equal to
$$R - \sqrt{(R^2 - x^2)}$$

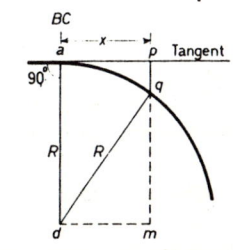

Proof: qm = $\sqrt{(R^2 - x^2)}$ pq = R - qm.
Alternatively, $pq = x^2/2R$, approximately.

Optical square method

Locate tangent points, also centre of circle and end of diameter opposite a, at e. With rods placed at a and e, move optical square as necessary to locate points such as x, y, z. Relies on angle in semicircle being 90°.
If fixed chord such as ax required, $ax = 2R\sin(\alpha/2)$, and for n even chord lengths, $\alpha = \Delta/n$.

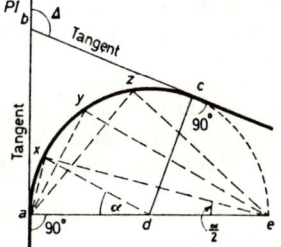

75

Chord bisection

If the ends of any chord of a circular arc are located, and the central angle of the chord and the arc radius are known, then the offset to the arc from the centre of the chord is equal to R x versine of half centre angle.

Principal application in replacement of missing pegs, as shown, but may be used to set out a complete curve by bisecting long chord then fixing crown point by offset R versine ($\Delta/2$), and continued bisection and offsets.

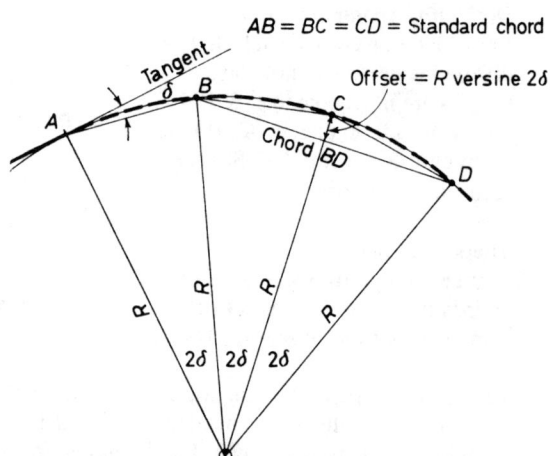

$AB = BC = CD$ = Standard chord

Offset = R versine 2δ

10·2   Transition curves

A transition curve has curvature which varies uniformly with distance along the curve. It is used to change curve radius and alter superelevation, either between circular curves of different radius or between a straight and a circular curve.

Figure illustrates vehicle on a canted section of carriageway. Gravity force W acts vertically, centrifugal force F is equal to Wv^2/gr, where v = vehicle velocity, g = standard gravity acceleration, and r = curve radius.

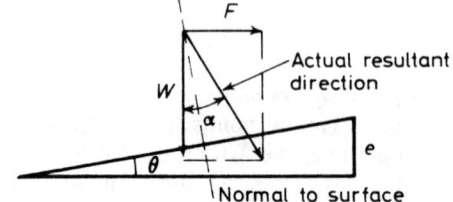

$F = Wv^2/gr$

e = cant

Resultant of F and W is normal to road surface under equilibrium conditions,

and $\tan\theta = Wv^2/Wgr = v^2/gr$ = centrifugal ratio.

Then $e = b\sin\theta \approx bv^2/gr$, since $\sin\theta \approx \tan\theta$ when θ is small.

∴ $e \propto 1/r$, and greater the radius the smaller the cant required

If cant applied uniformly along transition,

$e \propto l$, where l = distance along transition,

and $l \propto 1/r$, therefore lr is a constant.

Where transition meets circular arc, $l = L$ = transition length, and $r = R$ = circular arc radius, and $rl = RL$ Differentiating, and taking ϕ as angle consumed by the transition (deviation angle)

then $d\phi = dl/r$

∴ $d\phi/dl = 1/r = l/RL$

and integrating,

$\phi = l^2/2RL$, or $l = (2RL\,\phi)^{1/2}$

This is the Clothoid or Euler spiral equation, the ideal transition.

10·2·1   Minimum curve radius

Figure shows roadway with inadequate cant, and resultant of F and W at angle α to the vertical instead of angle θ.

Centrifugal ratio = $\tan\alpha = v^2/gr$. For roads, maximum value of $\tan\alpha$ is generally ¼ and maximum superelevation 1/10, then $\tan\theta/\tan\alpha = 0\cdot4$.

Minimum circular arc radius obtained from $v^2/gR = ¼$, or $R \nleq V^2/32$ metres, where V is the vehicle velocity in km/h.

76

10·2·2   Transition length

The usual criterion for deciding transition length is the rate of change of radial acceleration, q, expressed in metres per second cubed.

Vehicle velocity v, in metres per second, transition length L metres, time taken over transition is L/v seconds, and then maximum radial acceleration qL/v. Radial acceleration at junction of transition and circular arc is v^2/R,

then
$$qL/v = v^2/R,$$
$$\therefore \ L = v^3/qR \text{ metres.}$$

If velocity V km/h,

then
$$L = V^3/3 \cdot 6^3 \, qR.$$

Customary value for q is 1 ft/s^3, occasionally 2 ft/s^3.
Metric value 0·30 m/s^3, alternatives being 0·45 and 0·60 m/s^3.

10·2·3   Curves used for transitions

The ideal transition, the clothoid $l = (2RL \ \phi)^{1/2}$, gives rectilinear co-ordinates which are infinite series. Using x for distance along tangent to a point, and y for the perpendicular offset to the transition from the point,

$$x = l - l^5/10(2RL)^2 + l^9/216(2RL)^4 - \ldots$$
$$y = l^3/6RL - l^7/42(2RL)^3 + l^{11}/1320(2RL)^5 - \ldots$$

For practical use in metric units, refer to the tables published by the County Surveyor's Society.

Cubic spiral — If the first term of the series for y in the clothoid is used alone, the result is the cubic spiral with equation $y = l^3/6RL$. This may be set out by chords and offsets, with chords assumed equal to l in length, or alternatively by deflection angles.

Cubic parabola — If the first terms of the series for the clothoid are used, with $x = l$, the result is the cubic parabola with equation $y = x^3/6RL$. This is convenient for setting out by tangent distances and offsets.

Neither the cubic spiral nor parabola are ideal, but they are generally sufficiently good approximations in the absence of curve tables. The cubic parabola resembles the clothoid up to deviation angle of about 12°, but its radius of curvature increases again after deviation angle of about 24°.

Lemniscate — The equation of the lemniscate is $p = k \sqrt{\sin 2d}$, where p is chord length, k is a constant, and d is the deflection angle. This closely resembles the clothoid up to a polar angle of 45°. The curve is not widely used today.

10·2·4   Spiral and circular arc geometry

Angle Δ = deviation angle of the whole curve, T_1 and T_2 tangent points of the whole curve, O centre of the circular arc, as for simple circular curve, and R = radius of circular arc. T_3, T_4, junctions of transitions and circular arc.

AH, transition length, $= L$.

B$\hat{K}$H, deviation angle for transition, $= \phi_m$ = angle consumed by transition.

B$\hat{A}$H, deflection angle for transition, $= d_m = \frac{1}{3} \, \phi_m$, with sufficient accuracy for most purposes.

$d_m = \sin^{-1} GH/L = \sin^{-1} L^3/6RL^2 \approx L/6R$ radians $= 1800L/\pi R$ minutes.

$\phi_m = 3d_m \approx L/2R$ radians $= 3 \times 1800 \, L/\pi R$ minutes.

$\Delta - 2\phi_m$, angle consumed by circular arc.

77

HM, circular arc length equals
$\pi R(\Delta - 2\phi_m)/180$.
EH arc $\approx$ FH arc $= R\,\phi_m \approx L/2$.
AE $= L/2$.
AB, tangent length, equals
$L/2 + (R + S) \tan(\Delta/2)$.
S, shift, $= DF = L^2/6R - (R - R\cos\phi_m)$
$= L^2/24\,R$.
DE $= (L/2)^3/6RL = L^2/48R = S/2$.
$\widehat{KHA}$, back angle, $= \phi_m - d_m = 2d_m$.

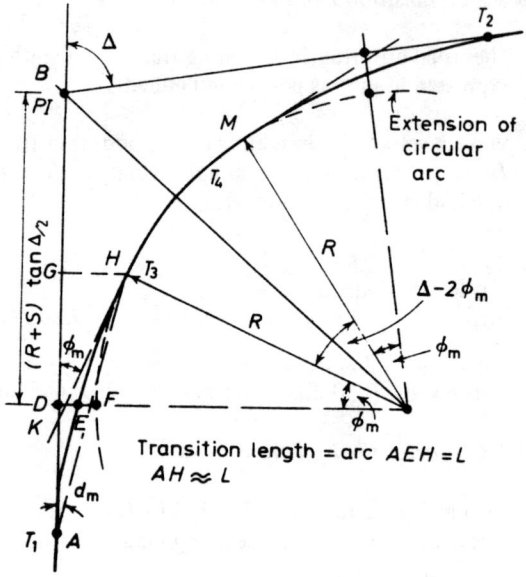

Extension of
circular
arc

Transition length = arc $AEH = L$
$AH \approx L$

Points on transition located by either of two
methods.

- (a) use $y = l^3/6RL$, where l = chord and d is its deflection angle
  $l^2/6RL$ radians $= 1800\,l^2/\pi RL$ minutes.
- (b) use $y = x^3/6RL$, where x is distance along tangent and y is the perpendicular offset from tangent.

10·2·5 Setting out the spiral and circular arc

- (a) Locate PI, measure Δ.
- (b) Fix radius R, compute transition length L and shift S.
- (c) Locate tangent points, with $T = L/2 + (R + S)\tan(\Delta/2)$.
- (d) Set out the first transition, either
  - (i) by theodolite deflection angles, with theodolite placed on A, T_1. Using $y = l^3/6RL$, compute deflection angles for chosen chord lengths $l_1, (l_1 + l_2), (l_1 + l_2 + l_3), \ldots$, deflection angles being
    $d_1 = \sin^{-1} l_1^2/6RL \approx l_1^2/6RL$ radians,
    $d_2 = \sin^{-1} (l_1 + l_2)^2/6RL \approx (l_1 + l_2)^2/6RL$ radians, etc., or
  - (ii) by tangent offsets,
    using $y = x^3/6RL$, the offset at any distance x along the tangent being $x^3/6RL$.
- (e) After completing transition, check that $GH = L^2/6R$.
- (f) Move theodolite to H, align on A, set off back angle $2d_m = L/3R$ radians, to define tangent to circular arc.
- (g) Set out circular arc in the usual way.

10·3 Vertical curves

Vertical curves are required at the intersection of differing road/rail gradients. Curves may be convex (summit) or concave (valley or sag). The requirements to be met by a vertical curve are:

- (a) constant change of gradient
- (b) uniform rate of increase of centrifugal force
- (c) adequate sighting distances.

The simple parabola is normally used, owing to its simplicity and constant gradient change. The cubic parabola is used on occasion.

Gradient sign convention: gradients rising to the right, positive, falling to the right negative. Specified as a percentage, e.g. $x\% = 1$ in $100/x$.
Left-hand gradient designated $p\%$, right-hand gradient $q\%$.

Grade angle = deflection angle = difference in % grade = $q\% - p\%$.
The correct signs of p and q must be inserted, according to particular case.

10·3·1 Properties of vertical curve parabola

Basic equation $y = ax^2$
$\therefore dy/dx = 2ax$
$\qquad = $ gradient of tangent to
$\qquad\quad$ curve.
$d^2y/dx^2 = 2a$
$\qquad\quad = $ constant,
the rate of change of gradient.

Change of gradient $= q\% - p\%$,
and in case shown $= -q - p$
$\qquad\qquad = -(p+q)$.

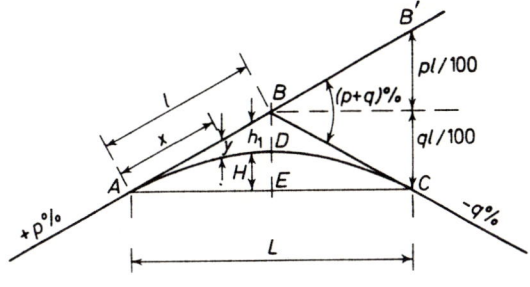

Since gradients involved generally small, it is usually sufficiently accurate to assume (a) perpendicular offsets from tangent equal to vertical offsets, (b) distance along tangent equal to horizontal distance, equal in turn to distance along the curve.

Difference in elevation between the curve and the tangent to it is equal to half the rate of change of gradient times the horizontal distance squared. (Distance measured horizontally from tangent point to point concerned on curve.) Then offset $y = ax^2$.

The horizontal lengths of any two tangents from a point to the vertical curve are equal, $\therefore$ AE = EC, or, the point of intersection of the two gradients is horizontally midway between the tangent points A and C.

Chord to a vertical curve has a rate of grade equal to that of the tangent at a point horizontally midway between the points of intercept of the chord.

The vertex of the parabola is at D, and BD = DE. Assuming the normal small gradients, the line BDE can be taken as being vertical.

Offset to curve from tangent — When $x = 2l$, $y = $ B'C $= (p+q)l/100 = ax^2 = a(2l)^2$
$$\therefore a = (p+q)/400l$$
and $\qquad\qquad\qquad y = (p+q)x^2/400l.$

Distance from intersection point to curve — When $x = l$, $y = $ BD $= (p+q)l/400 = $ DE.

Highest point of curve — Using A as datum for height, let H represent height of highest point of curve, and let this point be distant x from A,
then
$$H = xp/100 - (p+q)x^2/400l$$
and for H a maximum,
then
$$dH/dx = 0$$
$$\therefore \qquad x = 2pl/(p+q)$$
Curve length — $\qquad$ Since $a = (p+q)/400l$
$$\therefore \qquad l = (p+q)/400a$$
and total curve length
$$L = (p+q)/200a.$$

Since the parabola approximates to the circle, a limiting radius R is sometimes set. If limiting radius is R, then

$$d^2y/dx^2 = 2a$$
$$= 1/R,$$

$\therefore \quad 1/R = 2(p+q)/400\,l,$

$\therefore \quad R = 200l/(p+q),$

and minimum length based on radius is

$$L = R(p+q)/100.$$

10·3·2 Sight distances

Sight distance is the length of road over which an observer with eye level h metres above road surface can just see an object at h metres above the road surface on the opposite side of the road crest.

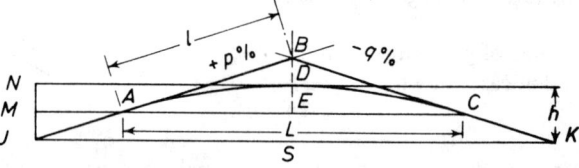

Sight line, as shown, is taken to pass tangentially through D, parallel to AC. Three cases arise
    (a)   sight distance = curve length.
    (b)   sight distance $>$ curve length.
    (c)   sight distance $<$ curve length.
Sight distance S, curve length $L =$ AC.

Case (a) $S = L$. Then $S = 2l$, and DE $= h = (p+q)l/400$. Given h, p, and q, then l can be found and $L = 800h/(p+q)$.

Case (b). $S > L$.
$$h = \text{JM} + \text{MN} = \text{JM} + \text{DE} = \text{JM} + \text{BD}.$$
$$\text{JM} = \left(\frac{p+q}{200}\right)\left(\frac{S}{2} - l\right)$$

and $\quad$ BD $= (p+q)l/400$

$\therefore \quad h = \left(\frac{S-l}{400}\right)(p+q)$

and $\quad L = 2S - 800h/(p+q)$.

Case (c). $S < L$. $\quad\quad$ Similarly, $L = S^2\,(p+q)/800h$.

10·3·3 Setting out vertical curves

    (a)   Determine gradients, compute curve length. p and q will always be known, but l, S, or a must be given.
    (b)   Fix chainages of tangent points and intersection, and their levels.
    (c)   Compute grade levels at required distances along tangents. Grade level = level of $A \pm px/100$, or level of $B \mp qx/100$.
    (d)   Compute offsets from tangent,
$$y = (p+q)x^2/400\,l.$$
    (e)   Compute curve levels, from
$$A \pm px/100 \mp (p+q)x^2/400\,l, \text{ or similarly from point B.}$$

Note: observe care in assigning signs to p and q and their difference.

The value fixed for h by M.O.T. is now 1·05 metres.